GWR Engines

Names, Numbers, Types & Classes

GWR Engines

Names, Numbers, Types & Classes

A reprint of the engine books
of 1911, 1928 and 1946
with some pages from that of 1938

DAVID & CHARLES : NEWTON ABBOT

0 7153 5367 5

These booklets were originally published
by Great Western Railway and
Great Western Railway Magazine

This composite edition published in 1971
by David & Charles (Publishers) Limited

Printed in Great Britain
by Redwood Press Limited Trowbridge Wilts
for David & Charles (Publishers) Limited
South Devon House Newton Abbot Devon

PUBLISHER'S NOTE TO THE 1971 EDITION

I was nine when the 1938 edition of GWR Engines: Names, Numbers, Types & Classes was published, and my shillingsworth of fact and picture had to last till I was seventeen, a quite different type of creature in a very different world. But it was the same GWR, and I remember the joy I experienced at being able to buy the 1946 edition, now 2s 6d but virtually unaltered, and so discard the tattered remains of the first that had kept me company on Newton Abbot station before the war, in the cellar during air raids, and on the first post-war forays to Paddington and Bristol.

Not only were the pre-war and post-war editions of the book virtually identical (a few 1938 features later dropped have been rescued for this omnibus work), but the continuity of locomotive practice on the GWR was such that many of the engines with lovely names included in the 1946 alphabetical list were already being listed back in 1911 when the publication of a book for engine spotters first began. Whenever one writes about the GWR one is trapped into commenting on its remarkable continuity. One class of locomotive, for instance, always seemed a natural development from the last, culminating in the famous 'Kings'; the American visit of the first 'King', *King George V*, is recorded in the 1928 edition included in this work. In 1946, of course, *King George V* still led the GWR's most powerful class and in fact had another decade of top-link service in front of it.

The GWR was innately conservative; what indeed was the *point* of replacing such a mighty class as the 'Kings' when they continued to give such fine service, whatever other railways and forms of transport might be doing? The GWR was proud, not infrequently triumphant, always knowing how to appeal to the public. Maybe it deserved to be Britain's most-loved railway; it certainly saw to it that it was. I have deliberately retained the preface to the 1938 edition and the advertisements at the end to demonstrate the pre-war attitude to enthusiasts. By 1946 nationalisation was already certain and it did not pay to be quite so fulsome.

Here, then, is the essence of GWR locomotive history, for those with technical interest, for entertainment or sheer nostalgia, or maybe just to show how a railway once 'sold' itself to 'boys of all ages'. The work is reproduced by arrangement with the Western Region's public relations officer, and I should also like to acknowledge thanks to Roger Burdett Wilson for his advice and loan of the copies from which this edition has been partly reproduced.

To end on a personal note, recently my firm published Mr Wilson's *Go Great Western: A History of GWR Publicity,* which lists all the editions of the locomotive and other books as well as giving a complete account of general publicity, posters, timetables and so on, with many examples reproduced. We published it from our offices over Newton Abbot station, offices which once housed the running and maintenance department controlling the largest single number of 'Kings' on the system. And it was under those very offices over 30 years ago that I first spotted engines, from 'Bulldogs' to 'Kings', decided that the GWR was the line for me, and rode the length of platform five aboard *King George V.* So for me this reprint seems the most natural thing in the world. To which I might add by way of a commercial that David & Charles have also published the two-volume 'Locomotive Monograph' on *The GWR Stars, Castles & Kings* and that we plan a similar work on *The GWR 4-4-0s.*

David St John Thomas

GREAT WESTERN RAILWAY.

NAMES

OF

ENGINES.

1911.

PUBLISHED BY
GREAT WESTERN RAILWAY MAGAZINE,
PADDINGTON STATION, LONDON, W.

CYLINDERS—Dia. 16″; Stroke, 18″.
BOILER—Barrel, 9′ 5″; Dia., 3′ 11″.

HEATING SURFACE—850 Sq. Ft.
WHEELS—Driving, 7′ 0″; Leading and Trailing, 4′ 0″.

GREAT WESTERN RAILWAY.

NAMES OF ENGINES.

Name.	No.	Type.	Name.	No.	Type.
Achilles	3031	4-2-2	Baden Powell ..	3374	4-4-0
Aden	3395	4-4-0	Badminton ..	3292	4-4-0
Agamemnon ..	3032	4-2-2	Barbados ..	3466	4-4-0
Albany	3456	4-4-0	Barrington ..	3293	4-4-0
Albatross ..	3033	4-2-2	Begonia	4102	4-4-0
Albert Brassey ..	3414	4-4-0	Bessborough ..	3295	4-4-0
Albert Edward ..	3062	4-2-2	Birkenhead ..	3443	4-4-0
Albion	171	4-6-0	Black Prince ..	3004	4-2-2
Alexander Hubbard	3299	4-4-0	Blasius	3353	4-4-0
Alfred Baldwin	3415	4-4-0	Blenheim ..	3294	4-4-0
Alliance	104	4-4-2	Bombay	3470	4-4-0
Amyas	3272	4-4-0	Bonaventura ..	3354	4-4-0
Anemone ..	4111	4-4-0	Boscawen ..	3254	4-4-0
Armorel	3273	4-4-0	Brasenose ..	3333	4-4-0
Armstrong ..	No. 7	4-4-0	Bride of Lammermoor	187	4-4-2
Atbara	3373	4-4-0	Brisbane ..	3396	4-4-0
Auckland ..	3393	4-4-0	Brunel	No. 16	4-4-0
Auricula	4101	4-4-0	Buffalo	1134	0-6-0 T
Australia.. ..	3455	4-4-0	Bulkeley ..	3072	4-2-2
Avalon	3332	4-4-0	Bulldog	3312	4-4-0

Cylinders—Four—Dia., 15″ ; Stroke, 26″.
Boiler—Barrel, 23′ 0″; Dia. outside, 5′ 6″ and 6′ 0″.
Heating Surface—3,400.81 Sq. Ft.

Wheels—Bogie, 3′ 2″; Driving and Intermediate, 6′ 8½″; Trailing, 3′ 8″.
Water Capacity of Tender, 3,500 Gallons.
Working Pressure—225 Lbs.

Name.	No.	Type.	Name.	No.	Type.
Blackbird ..	3731	4-4-0	City of Chester	3436	4-4-0
Bullfinch ..	3732	4-4-0	City of Gloucester	3437	4-4-0
			City of Hereford	3438	4-4-0
Cader Idris ..	1306	2-4-2 / T	City of London ..	3439	4-4-0
Calceolaria ..	4103	4-4-0	City of Truro ..	3440	4-4-0
Calcutta	3468	4-4-0	City of Winchester	3441	4-4-0
Calendula ..	4104	4-4-0	City of Exeter ..	3442	4-4-0
Cambria	3296	4-4-0	Colombo ..	3398	4-4-0
Camel	3352	4-4-0	Colonel Edgcumbe ..	3375	4-4-0
Camellia	4105	4-4-0	Columbia ..	3472	4-4-0
Camelot	3355	4-4-0	Comet	3315	4-4-0
Campanula ..	4106	4-4-0	Cornishman ..	3274	4-4-0
Cape Town ..	3397	4-4-0	Cornubia ..	3255	4-4-0
Cardiff	3444	4-4-0	Cotswold ..	3313	4-4-0
Carnation ..	4112	4-4-0	County of Middlesex ..	3473	4-4-0
Chancellor ..	154	2-4-0	County of Berks	3474	4-4-0
Charles Grey Mott	3417	4-4-0	County of Wilts..	3475	4-4-0
Charles Mortimer	3302	4-4-0	County of Dorset	3476	4-4-0
Charles Saunders	No. 14	4-4-0	County of Somerset ..	3477	4-4-0
Chepstow Castle	3314	4-4-0	County of Devon	3478	4-4-0
Chough	3275	4-4-0	County of Warwick ..	3479	4-4-0
Cineraria ..	4107	4-4-0	County of Stafford ..	3480	4 4-0
City of Bath ..	3433	4-4-0	County of Glamorgan ..	3481	4-4-0
City of Birmingham	3434	4-4-0			
City of Bristol ..	3435	4-4-0			

CYLINDERS—Four—Dia. $14\frac{1}{4}''$; Stroke 26″.
BOILER—Barrel, 14′ 10″; Dia. outside, 4′ $10\frac{13}{16}''$ and 5′ 6″.
HEATING SURFACE—2,014.4 Sq. Ft.

WHEELS—Bogie, 3′ 2″; Leading, Driving and Trailing, 6′ $8\frac{1}{2}''$.
WATER CAPACITY OF TENDER—3,500 Gallons.
WORKING PRESSURE—225 Lbs.

Name.	No.	Type.
County of Pembroke .	3482	4-4-0
County Carlow .	3801	4-4-0
County Clare ..	3802	4-4-0
County Cork ..	3803	4-4-0
County Dublin ..	3804	4-4-0
County Kerry ..	3805	4-4-0
County Kildare	3806	4-4-0
County Kilkenny	3807	4-4-0
County Limerick	3808	4-4-0
County Wexford	3809	4-4-0
County Wicklow	3810	4-4-0
County of Bucks	3811	4-4-0
County of Cardigan	3812	4-4-0
County of Carmarthen	3813	4-4-0
County of Chester	3814	4-4-0
County of Hants	3815	4-4-0
County of Leicester	3816	4-4-0
County of Monmouth ..	3817	4-4-0
County of Radnor	3818	4-4-0
County of Salop	3819	4-4-0
County of Worcester ..	3820	4-4-0
Cœur de Lion ..	180	4-4-2

Name.	No.	Type.
Courier	3006	4-2-2
Crusader ..	3036	4-2-2
Cyclops	No. 17	0-6-4 T
Chaffinch ..	3733	4-4-0
Cormorant ..	3734	4-4-0
Cheesewring ..	1311	0-6-0 T
Dartmouth ..	3356	4-4-0
Dartmoor	3276	4-4-0
David Mac Iver	3421	4-4-0
Dee	No. 71	2-4-0
Dominion of Canada ..	3453	4-4-0
Dog Star ..	4001	4-6-0
Dragon	3007	4-2-2
Dreadnought ..	3039	4-2-2
Duchess of Albany	3066	4-2-2
Duchess of Teck	3067	4-2-2
Duke of Cambridge ..	3068	4-2-2
Duke of Connaught ..	3065	4-2-2
Duke of Cornwall	3252	4-4-0
Duke of Edinburgh ..	3064	4-2-2
Duke of York ..	3063	4-2-2
Dunedin	3399	4-4-0

CYLINDERS—Dia. 18″; Stroke 30″.
BOILER—Barrel, 14′ 10″; Dia. outside, 4′ 10¾″ and 5′ 6″.
HEATING SURFACE—2,142.91 Sq. Ft.

WHEELS—Bogie, 3′ 2″; Leading, Driving and Trailing, 6′ 8½″.
WATER CAPACITY OF TENDER—3,500 Gallons.
WORKING PRESSURE—225 Lbs.

Name.	No.	Type.	Name.	No.	Type.
Durban	3400	4-4-0	Flamingo ..	3735	4-4-0
			Flying Dutchman	3009	4-2-2
Earl Cawdor ..	3297	4-4-0	Fowey	3281	4-4-0
Earl of Chester ..	3069	4-2-2	Fox	1391	0-4-0 / T
Earl of Cork ..	3418	4-4-0	Frank Bibby ..	3416	4-4-0
Earl of Devon ..	3277	4-4-0	Frederick Saunders ..	3042	4-2-2
Earl of Warwick	3070	4-2-2			
Eclipse	3334	4-4-0			
Eddystone ..	3278	4-4-0	Gardenia ..	4108	4-4-0
Edward VII ..	3413	4-4-0	Gibraltar.. ..	3401	4-4-0
Emlyn	3071	4-2-2	Glastonbury ..	3336	4-4-0
Emperor ..	3008	4-2-2	Godolphin ..	3358	4-4-0
Empire of India	3467	4-4-0	Goldfinch ..	3736	4-4-0
Empress of India	3040	4-2-2	Gooch	No. 8	4-4-0
Ernest Palmer ..	3420	4-4-0	Goonbarrow ..	1388	0-6-0 / T
Ernest Cunard ..	No. 98	4-6-0	Great Britain ..	3013	4-2-2
Etona	3335	4-4-0	Greyhound ..	3011	4-2-2
Eupatoria ..	3078	4-2-2	Grierson	3058	4-2-2
Evan Llewellyn	3419	4-4-0	Grosvenor ..	3298	4-4-0
Evening Star ..	4002	4-6-0	Guinevere ..	3257	4-4-0
Excalibur ..	3256	4-4-0	Guy Mannering	184	4-4-2
Exe	No. 72	2-4-0			
Exmoor	3279	4-4-0	Halifax	3402	4-4-0
			Hercules ..	3043	4-2-2
Fair Rosamund	1473	2-4-0 / T	Herschell ..	3376	4-4-0
Falmouth	3280	4-4-0	Hirondelle ..	3045	4-2-2

CYLINDERS—Dia. 18″; Stroke 30″.
BOILER—Barrel, 11′ 0″; Dia. outside, 5′ 0½″ and 4′ 5⅛″.
HEATING SURFACE—1,517.89 Sq. Ft.

WHEELS—Bogie, 3′ 2″; Driving and Trailing, 6′ 8½″; Radial Truck, 3′ 8″
WATER CAPACITY OF TANK—2,000 Gallons.
WORKING PRESSURE—195 Lbs.

Name.	No.	Type.	Name.	No.	Type.
Hobart	3403	4-4-0	Kimberley ..	3379	4-4-0
Holmwood ..	1813	0-6-0 / T	King Arthur ..	3258	4-4-0
Hotspur	3300	4-4-0	Kingsbridge ..	3359	4-4-0
Hyacinth ..	4113	4-4-0	Kirkland ..	178	4-6-0
			Kitchener ..	3377	4-4-0
Ilfracombe ..	3445	4-4-0	Knight of the Garter	4011	4-6-0
Isis	No. 73	2-4-0			
Isle of Jersey ..	3317	4-4-0	Knight of the Thistle..	4012	4-6-0
Isle of Guernsey	3316	4-4-0	Knight of St. Patrick	4013	4-6-0
Isle of Tresco ..	3288	4-4-0	Knight of the Bath	4014	4-6-0
Ivanhoe	181	4-4-2	Knight of St. John	4015	4-6-0
Jamaica	3464	4-4-0	Knight of the Golden Fleece	4016	4-6-0
John G. Griffiths	3060	4-2-2	Knight of the Black Eagle	4017	4-6-0
John Owen ..	1385	0-6-0 / T	Knight of the Grand Cross	4018	4-6-0
John W. Wilson..	3059	4-2-2	Knight Templar	4019	4-6-0
Jupiter	3318	4-4-0	Knight Commander	4020	4-6-0
Jackdaw ..	3737	4-4-0			
James Mason ..	3041	4-2-2	King Edward ..	4021	4-6-0
			King William ..	4022	4-6-0
Katerfelto ..	3319	4-4-0	King George ..	4023	4-6-0
Kekewich ..	3383	4-4-0	King James ..	4024	4-6-0
Kenilworth ..	3337	4-4-0	King Charles ..	4025	4-6-0
Khartoum ..	3378	4-4-0	King Richard ..	4026	4-6-0
Killarney ..	3408	4-4-0	King Henry ..	4027	4-6-0

CYLINDERS—Dia., 18″; Stroke, 26″.
BOILER—Barrel, 11′ 0″; Dia. outside, 4′ 10¾″ and 5′ 6″.
HEATING SURFACE—1,818.12 Sq. Ft.

WHEELS—Bogie, 3′ 8″; Driving and Trailing, 6′ 8½″.
WATER CAPACITY OF TENDER—3,000 Gallons.
WORKING PRESSURE—200 Lbs.

Name.	No.	Type.	Name.	No.	Type.
King John ..	4028	4-6-0	Lord Barrymore	174	4-6-0
King Stephen ..	4029	4-6-0	Lorna Doone ..	3047	4-2-2
King Harold ..	4030	4-6-0	Lyonesse ..	3361	4-4-0
Kingfisher ..	3738	4-4-0	Lyttelton ..	3404	4-4-0
Kilmar	1312	0-6-0 / T	Lady Margaret ..	1308	2-4-0 / T
Lady Superior ..	2901	4-6-0	Madras	3469	4-4-0
Lady of the Lake	2902	4-6-0	Mafeking.. ..	3382	4-4-0
Lady of Lyons ..	2903	4-6-0	Maine	3381	4-4-0
Lady Godiva ..	2904	4-6-0	Majestic	3048	4-2-2
Lady Macbeth ..	2905	4-6-0	Malta	3407	4-4-0
Lady of Lynn ..	2906	4-6-0	Marazion.. ..	3340	4-4-0
Lady Disdain ..	2907	4-6-0	Marco Polo ..	3339	4-4-0
Lady of Quality	2908	4-6-0	Marguerite ..	4114	4-4-0
Lady of Provence	2909	4-6-0	Marigold ..	4115	4-4-0
Lady of Shalott	2910	4-6-0	Maristow.. ..	3282	4-4-0
Ladysmith ..	3380	4-4-0	Marlborough ..	3303	4-4-0
La France ..	102	4-4-2	Mars	3341	4-4-0
Laira	3338	4-4-0	Mauritius ..	3405	4-4-0
Lalla Rookh ..	182	4-4-2	Melbourne ..	3406	4-4-0
Lambert ..	3055	4-2-2	Mendip	3323	4-4-0
Launceston ..	3360	4-4-0	Mercury	3321	4-4-0
Lightning ..	3016	4-2-2	Merlin	3260	4-4-0
Lobelia	4109	4-4-0	Mersey	3322	4-4-0
Lode Star ..	4003	4-6-0	Meteor	3320	4-4-0

CYLINDERS—Dia., 18″; Stroke, 30″.
BOILER—Barrel, 11′ 0″ ; Dia. outside, 5′ 0½″ and 4′ 5⅛″.
HEATING SURFACE—1,517.89 Sq. Ft.

WHEELS—Pony Truck, 3′ 2″; Intermediate & Driving, 5′ 8″; Radial [Truck, 3′ 8″.
WATER CAPACITY OF TANK—2,000 Gallons.
WORKING PRESSURE—195 Lbs.

Name.	No.	Type.	Name.	No.	Type.
Mignonette ..	4116	4-4-0	Pegasus	3343	4-4-0
Monarch ..	3301	4-4-0	Pembroke ..	3386	4-4-0
Montreal ..	3460	4-4-0	Pendennis Castle	3253	4-4-0
Morning Star ..	4004	4-6-0	Pendragon ..	3364	4-4-0
Mount Edgcumbe	3261	4-4-0	Penzance ..	3429	4-4-0
Mounts Bay ..	3283	4-4-0	Perseus	3345	4-4-0
			Petunia	4110	4-4-0
Natal Colony ..	3458	4-4-0	Peveril of the Peak ..	185	4-4-2
Narcissus ..	4117	4-4-0	Pluto	3344	4-4-0
Nelson	3049	4-2-2	Plymouth ..	3365	4-4-0
Newlyn	3362	4-4-0	Polyanthus ..	4118	4-4-0
Newport ..	3447	4-4-0	Powderham ..	3262	4-4-0
Newquay ..	3284	4-4-0	Powerful ..	3385	4-4-0
New Zealand ..	3454	4-4-0	President ..	103	4-4-2
North Star ..	No. 40	4-6-0	Pretoria	3389	4-4-0
Nightingale ..	3739	4-4-0	Primrose ..	4119	4-4-0
			Princess Beatrice	3076	4-2-2
			Princess Helena	3074	4-2-2
Omdurman ..	3384	4-4-0	Princess Louise ..	3075	4-2-2
One-and-All ..	3363	4-4-0	Princess May ..	3077	4-2-2
Orion	3342	4-4-0	Princess Royal ..	3073	4-2-2
Ottawa	3461	4-4-0	Peacock ..	3740	4-4-0
Oxford	3304	4-4-0	Pelican	3741	4-4-0
			Penguin ..	3742	4-4-0
Paddington ..	3448	4-4-0	Polar Star ..	4005	4-6-0

CYLINDERS—Dia., 18½″; Stroke, 30″.
BOILER—Barrel, 11′ 0″; Dia. outside, 4′ 10¾″ and 5′ 6″.
HEATING SURFACE—1,820.35 Sq. Ft.

WHEELS—Pony Truck, 3′ 2″; Coupled, 4′ 7½″
WATER CAPACITY OF TANK—1,800 Gallons.
WORKING PRESSURE—200 Lbs.

Name.	No.	Type.	Name.	No.	Type.
Quantock ..	3324	4-4-0	River Yealm ..	3432	4-4-0
Quebec	3409	4-4-0	Roberts	3387	4-4-0
Queensland ..	3471	4-4-0	Robertson ..	177	4-6-0
Quentin Durward	179	4-4-2	Robins Bolitho ..	173	4-6-0
Queen Mary ..	4031	4-6-0	Robin Hood ..	186	4-4-2
Queen Alexandra	4032	4-6-0	Rob Roy ..	188	4-4-2
Queen Victoria ..	4033	4-6-0	Royal Sovereign	3050	4-2-2
Queen Adelaide	4034	4-6-0	Royal Star ..	4008	4-6-0
Queen Charlotte	4035	4-6-0			
Queen Elizabeth	4036	4-6-0	Saint Agatha ..	2911	4-6-0
Queen Philippa ..	4037	4-6-0	Saint Ambrose ..	2912	4-6-0
Queen Berengaria	4038	4-6-0	Saint Andrew ..	2913	4-6-0
Queen Matilda ..	4039	4-6-0	Saint Augustine	2914	4-6-0
Queen Boadicea..	4040	4-6-0	Saint Bartholomew	2915	4-6-0
Racer	3018	4-2-2	Saint Benedict ..	2916	4-6-0
Reading	3449	4-4-0	Saint Bernard ..	2917	4-6-0
Red Gauntlet ..	183	4-4-2	Saint Catherine ..	2918	4-6-0
Red Star ..	4006	4-6-0	Saint Cuthbert ..	2919	4-6-0
Restormel ..	3366	4-4-0	Saint David ..	2920	4-6-0
Ringing Rock ..	1380	0-6-0 / T	Saint Dunstan ..	2921	4-6-0
Rising Star ..	4007	4-6-0	Saint Gabriel ..	2922	4-6-0
River Fal ..	3431	4-4-0	Saint George ..	2923	4-6-0
River Plym ..	3428	4-4-0	Saint Helena ..	2924	4-6-0
River Tamar ..	3268	4-4-0	Saint Martin ..	2925	4-6-0
River Tawe ..	3430	4-4-0	Saint Nicholas ..	2926	4-6-0

CYLINDERS—Dia., 19″; Stroke, 26″.
BOILER—Barrel, 11′ 0″; Dia. outside, 4′ 5½″ and 5′ 0½″.
HEATING SURFACE—1,517.89 Sq. Ft.

WHEELS—Bogie, 4′ 1½″; Driving and Trailing, 7′ 1½″.
WATER CAPACITY OF TENDER—3,000 Gallons.
WORKING PRESSURE—195 Lbs.

Name.	No.	Type.	Name.	No.	Type.
Saint Patrick ..	2927	4-6-0	Somerset ..	3327	4-4-0
Saint Sebastian ..	2928	4-6-0	Stanley Baldwin	3701	4-4-0
Saint Stephen ..	2929	4-6-0	Stephanotis ..	4120	4-4-0
Saint Vincent ..	2930	4-6-0	Steropes	No. 18	0-6-4 T
Samson	3305	4-4-0	Stormy Petrel ..	3051	4-2-2
Savernake ..	3308	4-4-0	Stour	No. 74	2-4-0
Sedgemoor ..	3351	4-4-0	St. Agnes ..	3287	4-4-0
Severn	3328	4-4-0	St. Anthony ..	3264	4-4-0
Shakespeare ..	3309	4-4-0	St. Aubyn ..	3367	4-4-0
Shelburne ..	3306	4-4-0	St. Austell ..	3326	4-4-0
Shrewsbury ..	3307	4-4-0	St. Columb ..	3325	4-4-0
Shooting Star ..	4009	4 6-0	St. Erth ..	3285	4-4-0
Singapore ..	3412	4-4-0	St. Germans ..	3265	4-4-0
Sir Francis Drake	3053	4-2-2	St. Ives	3266	4-4-0
Sir John Llewelyn	3422	4-4-0	St. Johns ..	3411	4-4-0
Sir Lancelot ..	3263	4-4-0	St. Just	3286	4-4-0
Sir Massey Lopes	3423	4-4-0	St. Michael ..	3267	4-4-0
Sir N. Kingscote	3424	4-4-0	Swallow	3023	4-2-2
Sir Redvers ..	3388	4-4-0	Swansea	3450	4-4-0
Sir Richard Grenville ..	3054	4-2-2	Swift	3356	4-4-0
			Swindon	3446	4-4-0
Sir Stafford ..	3368	4-4-0	Sydney	3410	4-4-0
Sir Walter Raleigh	3052	4-2-2	Seagull	3743	4-4-0
Sir Watkin Wynn	3427	4-4-0	Skylark	3744	4-4-0
Sir William Henry	3425	4 4-0	Starling	3745	4-4-0
Smeaton ..	3357	4-4-0			

Cylinders—Dia., 18″; Stroke, 30″
Boiler—Barrel, 14′ 10″; Dia. outside, 5′ 6″ and 4′ 10$\frac{13}{16}$″.
Heating Surface—2,142.91 Sq. Ft.

Wheels—Pony, 3′ 2″; Coupled, 4′ 7½″.
Water Capacity of Tender—3,000 Gallons.
Working Pressure—200 Lbs.

Name.	No.	Type.
Talisman	189	4-4-2
Tasmania	3457	4-4-0
Taunton	3451	4-4-0
Tavy	3346	4-4-0
Teign	No. 75	2-4-0
Terrible	3390	4-4-0
Thames	3329	4-4 0
The Abbot	172	4-4-2
The Great Bear	111	4-6-2
The Lizard	3259	4-4-0
The Wolf	3349	4-4-0
Thunderbolt	3079	4-2-2
Tintagel	3269	4-4-0
Titan	3348	4-4-0
Tor Bay	3290	4-4-0
Toronto	3459	4-4-0
Torquay	3372	4-4-0
Trefusis	3289	4-4-0
Tregeagle	3371	4-4-0
Tregenna	3291	4-4-0
Tregothman	3347	4-4-0
Trelawney	3369	4-4-0
Tremayne	3370	4-4-0
Tre Pol and Pen	3271	4-4-0
Trevithick	3270	4-4-0

Name.	No.	Type.
Trinidad	3465	4-4-0
Vancouver	3463	4-4-0
Viscount Churchill	175	4-6-0
Vulcan	3330	4-4-0
Walter Long	3426	4-4-0
Walter Robinson	3057	4-2-2
Waterford	3310	4-4-0
Waverley	190	4-4-2
Western Star	4010	4-6-0
Weymouth	3331	4-4-0
White	3392	4-4-0
Wilkinson	3056	4-2-2
William Dean	100	4-6-0
Will Scarlet	1356	0-6-0 T
Windsor Castle	3080	4-2-2
Winnipeg	3460	4-4-0
Winterstoke	176	4-6-0
Wolseley	3391	4-4-0
Wolverhampton	3452	4-4-0
Worcester	3027	4-2-2
Wye	No. 76	2-4-0
Wynnstay	3311	4-4-0

Cylinders—Dia., 18″; Stroke, 26″.
Boiler—Barrel, 11′ 0″; Dia. outside. 4′ 10¾″ and 5′ 6″.
Heating Surface—1,597.21 Sq. Ft.

Wheels—Coupled, 4′ 7½″; Truck, 2′ 8″.
Water Capacity of Tender—3,000 Gallons.
Working Pressure—200 Lbs.

CYLINDERS—Dia., 17″; Stroke, 24″.
BOILER—Barrel, 11′ 0″; Dia. outside, 4′ 5″ and 4′ 4⅛″
HEATING SURFACE—1,393.3 Sq. Ft.

WHEELS—5′ 2″ Diameter.
WATER CAPACITY OF TENDER—2,500 Gallons.
WORKING PRESSURE—150 Lbs.

Great Western Railway Engines

Names, Numbers, Types and Classes

(Price One Shilling)

PUBLISHED BY THE
GREAT WESTERN RAILWAY MAGAZINE,
PADDINGTON STATION, LONDON, W.

(1928)

_ C. W. R. LOCOMOTIVES _

_ REPRODUCED TO THE SAME SCALE _

"NORTH STAR" ———————— AS CONSTRUCTED BY R. STEPHENSON & CO IN 1837.
"LORD OF THE ISLES" ———— BUILT BY C. W. R. CO. AT SWINDON IN 1851.
"KINC CEORCE V" ———————— DO DO DO. DO. DO. 1927.

ENCINE	CYLINDERS		DRIVING WHEELS	BOILER PRESSURE	TRACTIVE EFFORT AT 85% BOILER PRESSURE
	Nº	DIMENSIONS			
"NORTH STAR"	2	16" x 16"	7 _ 0"	50 LBS.□"	2070 LBS
"LORD OF THE ISLES"	2	18" x 24"	8 _ 0"	140 "	9640 "
"KINC CEORCE V"	4	16¼" x 28	6 _ 6"	250 "	40300 "

Naming of Locomotives

On some of the railways of this country, naming locomotives is a very old custom. In the case of the Great Western, the practice has been identified with the Company's locomotive history in varying degree from the earliest periods.

Probably one would be near the mark in opining the first object in naming locomotives to have been a natural desire on the part of the makers to distinguish their productions from those of competitors when running in decisive trials, etc. Doubtless there were, and certainly are now, other reasons. For example, the naming of engines increases the interest of the public in particular types and in the trains they haul, and may, therefore, be looked upon as a means of advertisement. Again, the names selected often have an interesting connection and serve to perpetuate some important event, whilst in other instances, say those in which the names of persons, cities, towns, etc., are used, the selection of the name is intended as a compliment.

Reviewing the naming of engines of the Great Western Railway, it may be mentioned that the broad-gauge engines—passenger *and* goods—bore names, but not numbers. That there was method in appropriating the names is evident. Apparently the idea was to choose a "master" name, either mythical, historical, or physical, for each class, by which the particular type was identified, the individual engines of the class being given others arising out of, or appertaining to, the "master" name. Early examples of this method are given overleaf :—

Class.	Names.	Class.	Names.
" STAR " 1837–1841	North Star Morning Star Evening Star Dog Star Polar Star, etc.	" SUN " 1840–1841	Sun Sunbeam Meridian Eclipse Comet Meteor Aurora, etc.
" FIRE FLY " 1840	Fire Fly Spit Fire Wild Fire Fire Ball Fire King Fire Brand, etc.		

To give a further idea of the manner in which names were applied, and how they served the purpose of record, etc., we append a copy of the first two pages of the List of Broad-Gauge Passenger and Goods Locomotive Engines of the Great Western Railway in 1868.

DESCRIPTION OF THE CLASSES OF THE GREAT WESTERN RAILWAY LOCOMOTIVE ENGINES

Broad Gauge.

PASSENGER Engines in thick type, thus " **Abbot**."

" WOLF " .. .. Class, Tank Engines, Driving Wheels 6 ft. Single.

NOTE.—*The following of this Class have 7 ft. Driving Wheels, viz.: " Bright Star," " North Star," " Polar Star," " Red Star," " Rising Star," " Shooting Star," and " Orion."*

BOGIE .. .. Class, Tank Engines, Driving and Trailing Wheels 5 ft. 9 in. Coupled.

NOTE.—*" Brigand " and " Corsair " have 6 ft. Wheels Coupled.*

METROPOLITAN .. Class, Tank Engines, Driving and Trailing Wheels 6 ft. Coupled.

" PRIAM " .. .. Class, require Tenders, Driving Wheels 7 ft. Single.

NOTE.—*" Witch " has 7 ft. 6 in. Wheels.*

*NOTE. *" The Hawk," " Ostrich," " Phlegethon," " Pollux," Acheron," and " Cerberus," have been renewed as 6 ft. Coupled " Hawthorn " Class.*

" ALMA " .. .. Class, require Tenders, Driving Wheels 8 ft. Single.

" ABBOT " .. .. Class, require Tenders, Driving and Trailing Wheels 7 ft. Coupled.

" VICTORIA " .. Class, require Tenders, Driving and Trailing Wheels 6 ft. 6 in. Coupled.

"**HAWTHORN**" .. Class, require Tenders, Driving and Trailing Wheels 6 ft. Coupled.

GOODS **ENGINES** in italics, thus—" *Ajax*."

"*LEO*" .. .. Class, Tank Engines, Driving and Trailing Wheels 5 ft. Coupled.

BANKING .. .. Class, Tank Engines, all wheels 5 ft. Coupled.

"*FURY*" .. .. Class, require Tenders, all wheels 5 ft. Coupled (16 in. Cylinders).

"*CÆSAR*" .. .. Class, require Tenders, all wheels 5 ft. Coupled (17 in. Cylinders).

"*SWINDON*" .. Class, require Tenders, all wheels 5 ft. Coupled (17 in. Cylinders).

"*SIR WATKIN*" .. Metropolitan Class, Tank Engines, all wheels $4\frac{1}{2}$ ft. Coupled (17 in. Cylinders).

Name of Engine.	Maker.	Class.	Date of Starting.	Date when Renewed.
Abbot ..	Stephenson ..	Abbot ..	June '55	
Abdul Medjid	G.W.R. 5th lot ..	Victoria ..	Oct. '56	
Acheron ..	Fenton ..	Hawthorn ..	Jan. '42	Feb. '66
Achilles ..	Nasmyth ..	Priam ..	June '41	
Actæon ..	,, ..	,, ..	Dec. '41	Aug. '56
Æolus ..	Tayleur ..	Wolf ..	Nov. '37	
Ajax ..	G.W.R. 1st lot ..	Fury ..	May '46	
Alexander ..	,, 5th ,, ..	Victoria ..	Nov. '56	
Alligator ..	,, 2nd ,, ..	Cæsar ..	July '48	
Alma ..	Rothwell ..	Alma ..	Nov. '54	
Amazon ..	G.W.R. 4th lot ..	,, ..	Mar. '51	
Amphion ..	,, 6th ,, ..	Cæsar ..	April '56	
Antelope ..	Sharp	Wolf ..	Aug. '41	
Antiquary ..	Stephenson ..	Abbot ..	July '55	
Apollo ..	Tayleur ..	Wolf ..	Jan. '38	
Aquarius ..	Rothwell ..	Leo ..	June '42	
Arab ..	Rennie ..	Priam ..	April '41	
Argo ..	G.W.R. 1st lot ..	Fury ..	July '46	
Argus ..	Fenton ..	Priam ..	Aug. '42	July '64
Ariadne ..	G.W.R. 5th lot ..	Cæsar ..	Nov. '52	
Aries ..	Rothwell ..	Leo ..	June '41	
Arrow ..	Stothart ..	Priam ..	July '41	
Assagais ..	,, ..	Wolf ..	Sept. '41	Jan. '64
Atlas ..	Sharp	,, ..	June '38	July '60
Aurora ..	Hawthorn ..	,, ..	Dec. '40	
Avalanche ..	Stothart	*Banking* ..	Feb. '46	
Avon ..	G.W.R. 6th lot ..	Cæsar ..	June '57	
Avonside ..	Avonside Co. ..	Hawthorn ..	Dec. '65	
Azalia ..	G.W.R. 1st lot ..	*Metropolitan*	April '64	

Name of Engine.		Maker.	Class.	Date of Starting.	Date when Renewed.
Bacchus	..	G.W.R. 3rd lot	Fury ..	May '49	
Balaklava		Rothwell ..	Alma ..	Dec. '54	
Banshee	..	G.W.R. 6th lot	Cæsar ..	Sept. '54	
Bath	..	,, 12th ,, ..	Swindon ..	Jan. '66	
Bee	..	Vulcan Foundry	*Metropolitan*	July '62	
Behemoth	..	G.W.R. 2nd lot	Cæsar ..	Mar. '48	
Bellerophon		,, 1st ,,	Fury ..	July '46	
Bellona	..	Nasmyth ..	Priam ..	Nov. '41	
Bergion	..	G.W.R. 1st lot ..	Fury ..	Jan. '47	
Bey	..	Kitson ..	*Metropolitan*	July '62	
Beyer	..	Avonside Co. ..	Hawthorn	Dec. '65	

This shows that in 1868 there were eight " classes " of passenger and six of goods engines. Each class, it will be observed, following the practice referred to, was given a " master " name, the individual engines bearing other names. This system of only naming the engines continued with comparatively few exceptions throughout the period of the broad gauge, but with the change of gauge most of the named engines disappeared.

The " North Star " was constructed by Messrs. Robert Stephenson and Co. for the New Orleans Railway, America, but owing to a financial crisis at the time, was not delivered. Mr. Isambard Kingdom Brunel, who was then the Company's Engineer, arranged for its purchase for the Great Western Railway, the driving wheels being altered to 7′ 0″ diameter and the gauge modified to meet our requirements. It was the most powerful passenger locomotive of its day, and on June 1st, 1838, the first passenger train on the Great Western Railway was drawn by it from Paddington to Maidenhead at a speed of 36 m.p.h.

In the case of the narrow gauge, it was not the practice in the earlier years to name engines. The reason,probably was that in 1854 the Great Western Railway Company took over a batch of engines from the Shrewsbury and Chester Railway, bearing numbers 1 to 35. As these were the first 4 ft. 8 in. engines to come on to the line they were given similar Great Western numbers, and this started the numbering of narrow-gauge engines. Later,a further batch of engines obtained from the Shrewsbury and Birmingham Railway carried the numbers up to 56. A few of the engines bore name-plates at the time of their acquisition by the Great Western Railway Company, but the plates were gradually removed, probably to prevent confusion with the broad-gauge engines. The first Great Western narrow-gauge engine built at Swindon came out the following year, 1855. It was numbered 57, and from this time the *numbering* of Great Western engines became the standard practice.

Among the earlier of the narrow-gauge engines to be named, in addition to being numbered, were some of the famous 7 ft. singles, including the following :—

No.	Date constructed.	Name.
378	Sept. 1866	Sir Daniel
380	,, ,,	North Star
381	Oct. ,,	Morning Star
471	June 1869	Sir Watkin
55	Sept. 1873	Queen
999	Mar. 1875	Sir Alexander
1118	,, ,,	Prince Christian
1122	April ,,	Beaconsfield
1123	May ,,	Salisbury

With these comparatively few exceptions, the *naming* of narrow-gauge engines, up to 1892, was not resorted to.

In 1892—the year of the conversion of the gauge—the first of the " 7 ft. 8 in. singles " was put into service. These were originally " convertibles," bearing numbers only, but after conversion, names were given them. Simultaneously four four-coupled " convertibles " were converted, and were named " Armstrong," " Gooch," " Brunel," and " Charles Saunders."

A little later the number of 7 ft. 8 in. singles was increased. Each was named, and hereabouts the *naming*, in addition to the *numbering*, of the passenger engines came more into vogue. The fleet of " 7.8's " steadily grew, and in choosing names for them many of the old broad-gauge "8 ft." names, such as "Amazon," "Courier," "Rover," "Sultan," were perpetuated.

In 1895 the first of the 4–4–0 type, for working over the heavy inclines in the West country, were constructed. By reason of their "vocation" they were, as a class, termed "Devons." The first, however, was named "Duke of Cornwall," and the class eventually became known as "Dukes."

In 1897 the "Badmintons," 4–4–0 inside cylinders, were put into service. The name "Badminton" was chosen after the celebrated Hunt, and the engines have always been referred to as "Badmintons."

The following year, No. 3311—"Bulldog"—appeared, getting its name from the impression of strength it gave, compared with other engines.

Later came the "Atbaras," a development of the "Badmintons," with straight frames. The names appropriated to them are identified with outstanding incidents with which a certain celebrated soldier of the Empire was closely associated..

1902 witnessed the first of the 4–6–0 2-cylinder type, which, after running some months, was named "William Dean" (2900).

The famous "Cities" were introduced the following year, and from this time may be said to have commenced a system of naming successive batches and types of engines with the view, first, of securing ready identification, and, later on, of helping in a scheme of standardisation which it was desired to accomplish. The names appropriated to the "Cities" were chosen after cathedral and other cities through which the Company's line runs; moreover, the common word "City" indicated a class.

The year 1903 witnessed the advent of the De Glehn compound, to which was given the name "La France." Later, two more powerful engines of this type were purchased, and appropriate names—"President" and "Alliance"—were chosen for them, their acquisition synchronising with the launching of "L'Entente Cordiale."

In 1905, six 2-cylinder 4–6–0's, similar to the sample one built in 1902, were constructed. Five were eventually given the names of certain members of the Great Western Railway directorate. Contemporaneously, an entirely new type (4–4–2—" Atlantics "—outside cylinders) appeared. It was decided to appropriate to these, names borne by the broad-gauge " Lalla Rookh " class (4–4–0—1855), viz., " Ivanhoe," " Lalla Rookh," " Redgauntlet," " Robin Hood," " Rob Roy," " The Abbot," " Waverley." Later they were converted to 4–6–0's.

The year 1905 saw also the appearance of the " Counties." Possessing features peculiar to the class, the plan followed in the case of the " Cities " was repeated, the names being selected from the counties served by the railway. The first was " County of Middlesex," the idea being to start at Paddington, which is in Middlesex, and proceed down the line through Berkshire, Wiltshire, etc., naming an engine after each county. Later on, the type having to be increased to a greater number than there were counties served by the Great Western Railway, other counties were selected.

In 1907 the first of the 4-cylinder engines (No. 40—now 4000) came into service. Having regard to the engine's exceptional design, it was decided to perpetuate upon it the name (" North Star ") borne by two famous predecessors, firstly the old broad-gauge and later the narrow-gauge 7 ft. Subsequent engines of this type were also given names in succession to the broad-gauge " Star " series.

Next came a batch of 2-cylinder 4–6–0's. Hereabouts, so far as engine nomenclature was concerned, it became the practice to employ, as in the case of the " Cities " and " Counties," names made up either of two or three words, the idea being that either the first or last word should be common to the type, and the remainder of the name lend itself to alphabetical reference. Thus it came about that this batch of 4–6–0's were given names commencing with " Lady " in commemoration of certain famous characters in fiction. The following year a further number of engines of this type were built, possessing one or two distinctive features, to which the names of " Saints " were appropriated.

A year later another batch of engines similar to the

" Stars " were put into service and given names indicative of Orders of Knighthood.

A new type of 6 ft. 8½ in. 4–4–0 inside cylinder engines was introduced in 1909, to which were given the names of flowers, thus individualising the class. This type was followed by a " lot " of 5 ft. 8 in. 4–4–0 inside cylinder engines, to which were appropriated the names of birds, the " Bird " class signalising a temporary return to the 5 ft. 8 in. inside cylinder type. The year 1909 saw also the appearance of more of the 4–6–0, 4-cylinder engines similar to the " Stars " and " Knights," and it was decided to name these after certain of the Kings of England. The first was " King Edward," its availability synchronising with a notable public event with which His late Majesty was identified. Two years later a further batch came out, and, appropriately, were named after Queens of England, " Queen Mary " being the first.

In 1911 an additional number of " Saints " and " Ladies" were constructed. They were named after historical residences, and became known as the " Courts."

The year 1913 brought more " Stars." Commencing with " Prince of Wales," they were named after the princes of the reigning Royal Family.

From the foregoing will be seen that during the earlier history of the narrow-gauge engines, *naming* was only occasionally resorted to, but that in later years the appropriation of names has proceeded upon an arranged basis, one effect being that the engines, by the names selected for them, have been brought into classes as follows :—

Inside Cylinders.

4-4-0 Type. 5′ 8″ wheels.		4-4-0 Type. 6′ 8½″ wheels.	
	Class.		Class.
	" DUKE "		" CITY "
Bulldog ... ... Bird	" BULLDOG "	Badminton ... ... Atbara... ... ... Flower...	" FLOWER "

Outside Cylinders.				Four Cylinders.			
4-4-0 Type. 6′ 8½″ wheels.	4-6-0 Type. 6′ 8½″ wheels		4-6-0 Type. 6′ 0″ wheels	4-6-0 Type. 6′ 8½″ wheels.		4-6-0 Type. 6′6″ wheels.	
Class.		Class.	Class		Class.		Class.
" COUNTY"	Scott ... Lady ... Saint ... Court ...	" SAINT "	" SAINT MARTIN"	Star ... Knight Monarch Queen Prince Princess Abbey	" STAR "		" KING "
					"CASTLE"		

The naming, in addition to the numbering of engines is not really necessary. There is, however, a good deal to be said in favour of it, one valuable feature being that it serves as an aid to memory. Numbers are difficult to carry in mind, and it may fairly be said to be the case, so far as Great Western engines are concerned, that the names of individuals and of types constitute a means of reference of a distinctly useful character in many directions.

These notes are based upon what may be termed the official naming of the locomotives. It should not be left unmentioned that engines which are not named officially are, so to speak, by custom almost, given descriptive appellations directly they go into service by which, peculiarly they are during the whole of their " life " referred to and identified. Instances of this custom are the " Aberdares" (2–6–0 goods), built for the Aberdare coal service ; the " County Tanks," and so on—a further indication of the usefulness of naming engines.

Lastly, there is the now almost universal practice of identifying types by the wheel arrangement, viz., " 4–6–0," " 4–4–2T," etc.

It will not be difficult to appreciate that the standardising of engines, including the parts thereof, together with the boilers, as far as practicable, lends itself to nomenclature. At this juncture, therefore, we may, appropriately in chronological sequence, outline—with the view of coupling up with that mentioned in the foregoing pages as to the naming of the engines—what has been done in this direction at the Swindon Works in recent years.

1895 First engine of the 4–4–0 type, with 5 ft. 8 in. driving wheels (and the now well-known extended smoke box) built for service in South Devon. Proving successful, more of the type were constructed later.

1896 First 4–6–0 type engine, No. 36, with 4 ft. 7½ in. driving wheels and 20 in. × 24 in. cylinders, built for heavy goods work

1899 A modification of this design produced the " Krugers," which had 4 ft. 7½ in. driving wheels, and cylinders 19 in. diameter by 28 in. stroke fitted with piston valves.

1897 The " Badmintons," forerunners of the " Cities," were produced, having the same cylinders and motion as the " Devonshires " or " Dukes."

1898 to 1900 Fleet of " Devonshires " augmented. The last 20 (" Bulldogs ") were fitted with the standard No. 2 boiler, and instead of the outside frames being cut down between the driving and trailing wheels they were continued straight through. In this form, and with the exception that the boiler barrel is coned, a large number of these engines have since been built.

1900 to 1901 The " Atbaras," a further development of the " Badmintons," were turned out, being somewhat similar to the " Bulldogs " and having standard No. 2 boilers.

1900 First of the " Aberdares," 2–6–0 class (No. 33), built. Successor to the " Krugers," but having coupled wheels 4 ft. 7½ in. diameter. The following year more of these were built to the same drawings, and since that time the number has been increased, but they have now been fitted with No. 4 boilers.

1902 First engine of the 4–6–0 type 2-cylinder engines (No. 100—now 2900), was built. The boiler had a straight barrel, but on succeeding engines the barrel was coned.

1903 First of the 2–8–0 type (No. 97—now 2800) goods engines was built. In September the first of the 2–6–2T class (No. 99—now 3100) was put into traffic. 4–4–2 De Glehn compound locomotive purchased.

1904 The first of the "Counties" (No. 3473—now 3800) 4–4–0 type was built.

1905 The "County Tanks" were introduced, 4–4–2T type. Two more De Glehn compounds of greater power were acquired. Although not Great Western standard, they were subsequently brought within the "interchangeability—of-boiler" scheme, and provided with Great Western boilers.

1907 The first of the 4-cylinder engines built (No. 40—now 4000). It was built experimentally with a 4–4–2 wheel arrangement, but later altered to 4–6–0 type, many of which are now in service.

1908 In June, the first engine of the "Pacific" type (4–6–2) built in this country, viz. No. 111—"The Great Bear"—was completed. This engine was a development of the "Stars," the cylinders, motion, expansion gear, axles and bogie being similar.

1910 First of the 2-6-0 outside cylinder engines produced, an important feature being the interchangeability of parts with the 2-6–2T type.

1910 First engine of the 2-8-0T type produced, followed by 2-8-0 (4700 class) for dealing with heavy express mixed traffic. The latter was introduced in May, 1919, and designed to carry a standard No. 7 boiler. This engine has the same wheel arrangement as the standard 28XX class, but instead of $55\frac{1}{2}$ in. driving wheels has wheels 68 in. diameter.

1910 to 1919 Standard types of engines turned out, there being no exceptional development owing to the War.

1923 In September the "Castle" class of engine was put into service. This class of engine having proved very successful, it was decided to build more of them. In addition, four of the older 4-cylinder engines, viz., "Shooting Star," "Knight of the Golden Fleece," "Queen Alexandra" and "Queen Philippa" were converted to this class.

1924 "The Great Bear"—No. 111—the only Great Western engine of the "Pacific" type was reconstructed as a "Castle" and re-named "Viscount Churchill." In December, a new type of engine—0-6-2T—(No. 5600) specially designed to deal with the heavy traffic of the South Wales coal fields, over the long grades and sharp curves met with in that area, was put into service.

1925 One of the 4-6-0 class, No. 2925 "Saint Martin," was rebuilt with its coupled wheels 6 ft. 0 in. in diameter instead of 6 ft. $8\frac{1}{2}$ in., and fitted with a cab similar to the "Castle" class.

1927 In July the "King" class of engine was put into service. These engines, the most powerful of the passenger type in Great Britain, were specially designed to cope with the heavy traffic and fast running on certain sections of the line, and embody several special features. The first,

" King George V," was shipped to America and exhibited at the Baltimore and Ohio Railroad Centenary. It was accompanied by the Broad Gauge engine " North Star," reconstructed (as it was built in 1837) at Swindon Works in connection with the Railway Centenary celebrations in this country in 1925.

" Star " class engines bearing the names of " Kings " were re-named " Monarchs." The nameplates were also removed from other odd engines bearing the names of Kings.

1928 Additional engines of the " King " class, and a new series of the " Saint Martin " type are in course of construction, the latter to be named after well known Halls on the G.W.R. system.

A.J.L.W.

Serial Numbers :—

3252 — 3291.

3252 Class. Type 4—4—0.

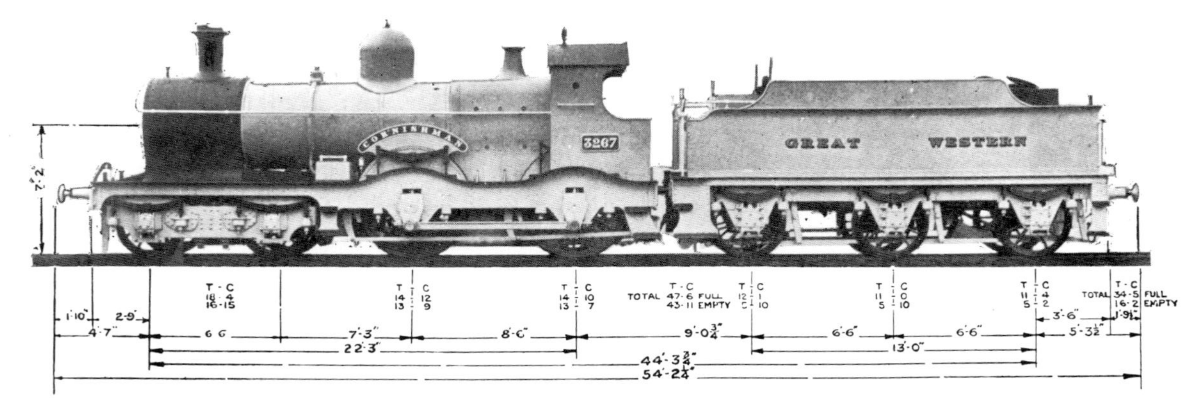

CYLINDERS—Diam. 18″; Stroke. 26″;
BOILER—Barrel, 11′ 0″; Diam. Outs., 4′ 5″ and 4′ 4″.
FIREBOX—Outs, 5′ 10″ by 4′ 0″; Ins., 5′ $1\frac{7}{16}$″ by 3′ 4″; Height, 6′ 0½″
TUBES—Superheater Tubes, No. 36: Diam., 1″; Length, 11′ 1⅜″. Fire Tubes, No. 195: Diam., 1⅝″: Length, 11′ 3 11/16″ Fire Tubes, No.6 Diam., 5⅛″; Length, 11′ 3 11/16″.

HEATING SURFACE—Superheater Tubes, 81.25 sq. ft. Fire Tubes 1028.95 sq. ft. Firebox 113.95 sq. ft. Total, 1,224.15 sq. ft.,
AREA OF FIREGRATE—17.2 sq. ft.
WHEELS—Bogie, 3′ 8″; Coupled, 5′ 8″.
WATER CAPACITY OF TENDER—2,500 gallons.
WORKING PRESSURE—180 lbs.
TRACTIVE EFFORT—18 955 lbs.

GREAT WESTERN RAILWAY.

Named Engines.

" DUKE CLASS "

No.	Name.	No.	Name.
3252	Duke of Cornwall	3272	Fowey
3253	Boscawen	3273	Mounts Bay
3254	Cornubia	3274	Newquay
3255	Excalibur	3275	St. Erth
3256	Guinevere	3276	St. Agnes
3258	The Lizard	3277	Isle of Tresco
3259	Merlin	3278	Trefusis
3260	Mount Edgcumbe	3279	Tor Bay
3261	St. Germans	3280	Tregenna
3262	St. Ives.	3281	Cotswold
3263	St. Michael	3283	Comet
3264	Trevithick	3284	Isle of Jersey
3265	Tre Pol and Pen	3285	Katerfelto
3266	Amyas	3286	Meteor
3267	Cornishman	3287	Mercury
3268	Chough	3288	Mendip
3269	Dartmoor	3289	St. Austell
3270	Earl of Devon	3290	Severn
3271	Eddystone	3291	Thames

Serial Numbers :—

3300 — 3455.

3300 Class. Type 4—4—0.

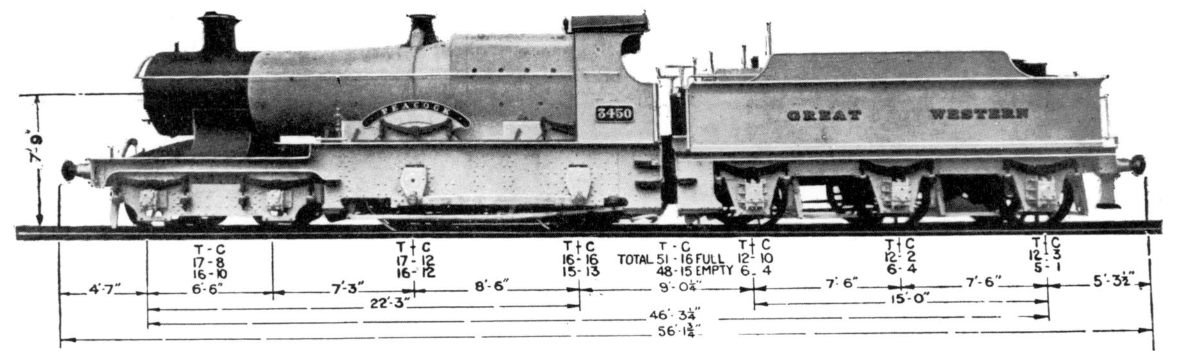

CYLINDERS—Diam., 18" ; Stroke, 26".
BOILER—Barrel, 11' 0" ; Diam. Outs. 4' 5⅛" and 5' 0½".
FIREBOX—Outs., 7' 0" by 5' 3" & 4' 0" ; Ins., 6' 2 11/16" by 4' 3½" & 3' 2¾" ; Height, 6' 0 7/16" and 5' 0 7/16".
TUBES—Superheater Tubes, No. 36 : Diam., 1" ; Length, 11' 5⅜". Fire Tubes, No. 218 : Diam., 1⅝" ; Length, 11' 4 5/16". Fire Tubes No. 6 ; Diam. 5⅛" ; Length, 11' 4 5/16".

TOTAL WEIGHT OF ENGINE—51 tons 16 cwt Full.
48 ,, 15 ,, Empty

HEATING SURFACE—Superheater Tubes, 82.20 sq. ft. Fire Tubes. 1,144·94 sq. ft. Firebox 121·80 sq. ft. Total, 1,348·95 sq. ft.
AREA OF FIREGRATE—20·35 sq. ft.
WHEELS—Bogie, 3' 8" ; Coupled, 5' 8".
WATER CAPACITY OF TENDER—3,000 gallons.
WORKING PRESSURE—200 lbs.
TRACTIVE EFFORT—21,060 lbs.

TOTAL WEIGHT OF TENDER—36 tons 15 cwt Full.
17 ,, 9 ,, Empty.

"BULLDOG" CLASS

No.	Name.	No.	Name.	No.	Name.
3302	Sir Lancelot	3347	Kingsbridge	3398	Montreal
3303	St. Anthony	3348	Launceston	3399	Ottawa
3304	River Tamar	3349	Lyonesse	3400	Winnipeg
3305	Tintagel	3350	Newlyn	3401	Vancouver
3306	Armorel	3351	One-and-All	3402	Jamaica
3307	Exmoor	3352	Pendragon	3403	Trinidad
3308	Falmouth	3353	Pershore Plum	3404	Barbados
3309	Maristow	3354	Restormel	3405	Empire of India
3310	St. Just	3355	St. Aubyn	3406	Calcutta
3311	Bulldog	3356	Sir Stafford	3407	Madras
3312	Isle of Guernsey	3357	Trelawny	3408	Bombay
3313	Jupiter	3358	Tremayne	3409	Queensland
3314	Mersey	3359	Tregeagle	3410	Columbia
3315	Quantock	3360	Torquay	3411	Stanley Baldwin
3316	St. Columb	3362	Albert Brassey	3412	John G. Griffiths
3317	Somerset	3363	Alfred Baldwin	3413	James Mason
3318	Vulcan	3364	Frank Bibby	3414	A. H. Mills
3319	Weymouth	3365	Charles Grey	3415	George A. Wills
3320	Avalon		Mott	3416	John W. Wilson
3321	Brasenose	3366	Earl of Cork	3417	Lord Mildmay of
3322	Eclipse	3367	Evan Llewellyn		Flete
3323	Etona	3369	David MacIver	3418	Sir Arthur Yorke
3324	Glastonbury	3370	Sir John	3422	Aberystwyth
3325	Kenilworth		Llewelyn	3430	Inchcape
3326	Laira	3371	Sir Massey Lopes	3434	Joseph Shaw
3327	Marco Polo	3372	Sir M. Kingscote	3439	Weston-Super-
3328	Marazion	3373	Sir William		Mare
3329	Mars		Henry	3441	Blackbird
3330	Orion	3374	Walter Long	3442	Bullfinch
3331	Pegasus	3375	Sir Watkin Wynn	3443	Chaffinch
3332	Pluto	3376	River Plym	3444	Cormorant
3333	Perseus	3377	Penzance	3445	Flamingo
3334	Tavy	3378	River Tawe	3446	Goldfinch
3335	Tregothnan	3379	River Fal	3447	Jackdaw
3336	Titan	3380	River Yealm	3448	Kingfisher
3337	The Wolf	3381	Birkenhead	3449	Nightingale
3338	Swift	3383	Ilfracombe	3450	Peacock
3339	Sedgemoor	3391	Dominion of	3451	Pelican
3340	Camel		Canada	3452	Penguin
3341	Blasius	3392	New Zealand	3453	Seagull
3342	Bonaventura	3393	Australia	3454	Skylark
3343	Camelot	3394	Albany	3455	Starling
3344	Dartmouth	3395	Tasmania		
3345	Smeaton	3396	Natal Colony		
3346	Godolphin	3397	Toronto		

Serial Numbers :—

4100 — 4172.

4100 Class. Type 4—4—0.

CYLINDERS—Diam., 18″ ; Stroke, 26″.
BOILDER—Barrel, 11′ 0″ ; Diam. Outs., 4′ 5¼″ and 5′ 0½″.
FIREBOX—Outs., 7′ 0″ by 5′ 3″ and 4′ 0″ ; Ins., 6′ 2 11/16″ by 4′ 3½″ and 3′ 2¾″ Height, 6′ 0 7/16″ and 5′ 0 7/16″.
TUBES—Superheater Tubes, No. 36 : Diam., 1″ ; Length, 11′5⅜″. Fire Tubes, No. 218 ; Diam., 1⅝″ ; Length, 11′ 4 5/16″. Fire Tubes No. 6: Diam., 5⅛″ ; Length, 11′ 4 5/16″.

TOTAL WEIGHT OF ENGINE—53 tons 6 cwt. Full.
49 ,, 18 ,, Empty.

HEATING SURFACE—Superheater Tubes, 82·20 sq. ft., Fire Tubes 1,144·95 sq. ft. Firebox, 121·80 sq. ft. Total 1,348·95 sq ft.
AREA OF FIREGRATE—20·35 sq. ft.
WHEELS—Bogie, 3′ 8″ ; Coupled, 6′ 8½″.
WATER CAPACITY OF TENDER, 3,500 gallons.
WORKING PRESSURE—200 lbs.
TRACTIVE EFFORT—17,790 lbs.

TOTAL WEIGHT OF TENDER—40 tons 0 cwt. Full.
18 ,, 5 ,, Empty.

"FLOWER" CLASS.

No.	Name.	No.	Name.
4100	Badminton	4141	Aden
4101	Barrington	4142	Brisbane
4102	Blenheim	4143	Cape Town
4103	Bessborough	4145	Dunedin
4104	Cambria	4148	Singapore
4106	Grosvenor	4149	Auricula
4107	Alexander Hubbard	4150	Begonia
4108	Hotspur	4152	Calendula
4109	Monarch	4154	Campanula
4110	Charles Mortimer	4156	Gardenia
4111	Marlborough	4157	Lobelia
4113	Samson	4158	Petunia
4120	Atbara	4159	Anemone
4121	Baden Powell	4161	Hyacinth
4122	Colonel Edgcumbe	4162	Marguerite
4124	Kitchener	4163	Marigold
4127	Ladysmith	4164	Mignonette
4129	Kekewich	4167	Primrose
4130	Omdurman	4168	Stephanotis
4131	Powerful	4169	Brunel
4132	Pembroke	4170	Charles Saunders
4137	Wolseley	4171	Armstrong
4138	White	4172	Gooch
4139	Auckland		

Serial Numbers :—

3700 — 3719.

3700 Class. Type 4—4—0.

CYLINDERS—Diam., 18″; Stroke 26″.
BOILER—Barrel, 11′ 0″; Diam. Outs., 4′ 10¾″ and 5′ 6″.
FIREBOX—Outs. 7′ 0″ by 5′ 9″ and 4′ 0″; Ins., 6′ 2½″ by 4′ 8⅞″ and 3′ 2⅝″, Height, 6′ 6⅜″ and 5′ 0⅜.
TUBES—Superheater Tubes, No. 84: Diam., 1″; Length, 11′ 5⅜″. Fire Tubes, No. 235: Diam., 1⅝″; Length, 11′ 4 7/16″. Fire Tubes No. 14; Diam., 5⅛″; Length, 11′ 4 7/16″.

TOTAL WEIGHT OF ENGINE—55 tons 6 cwt. Full.
50 ,, 19 ,, Empty.

HEATING SURFACE—Superheater Tubes, 191·79 sq. ft. Fire Tubes 1,349·64 sq. ft. Firebox, 128·72 sq. ft. Total, 1,670·15 sq. ft.
AREA OF FIREGRATE—20·56 sq. ft.
WHEELS—Bogie, 3′ 8″; Coupled, 6′ 8½″.
WATER CAPACITY OF TENDER—3,500 gallons.
WORKING PRESSURE—200 lbs.
TRACTIVE EFFORT—17,790 lbs.

TOTAL WEIGHT OF TENDER—40 tons 0 cwt. Full.
18 ,, 5 ,, Empty.

"CITY" CLASS.

No.	Name.	No.	Name.
3700	Durban	3710	City of Bath
3701	Gibraltar	3711	City of Birmingham
3702	Halifax	3712	City of Bristol
3703	Hobart	3713	City of Chester
3704	Lyttelton	3714	City of Gloucester
3705	Mauritius	3715	City of Hereford
3706	Melbourne	3716	City of London
3707	Malta	3717	City of Truro
3708	Killarney	3719	City of Exeter
3709	Quebec		

Serial Numbers :—

2900 — 2924.
2926 — 2955.
2971 — 2990.
2998.

2900 Class. Type 4—6—0.

CYLINDERS—Diam., 18½"; Stroke, 30".
BOILER—Barrel, 14' 10"; Diam. Out., 5' 6" and 4' 10 13/16".
FIREBOX—Outs., 9' 0" by 5' 9" and 4' 0"; Ins., 8' 2 7/16" by 4' 9" and 3' 2⅝"; Height, 6' 6⅜" and 5' 0⅜".
TUBES—Superheater Tubes, No. 84: Diam., 1"; Length, 15' 3⅜". Fire Tubes, No 176: Diam., 2"; No. 14: Diam., 5⅛"; Length, 15' 27/16".

TOTAL WEIGHT OF ENGINE—72 tons 0 cwt. Full.
66 ,, 0 ,, Empty.

HEATING SURFACE—Superheater Tubes, 262·62 sq ft. Fire Tubes 1,686 60 sq. ft. Firebox 154·78 sq. ft. Total, 2,104·00 sq. ft.
AREA OF FIREGRATE—27·07 sq. ft.
WHEELS—Bogie, 3' 2"; Coupled, 6' 8½".
WATER CAPACITY OF TENDER—3, 500 gallons.
WORKING PRESSURE—225 lbs.
TRACTIVE EFFORT—24,395 lbs.

TOTAL WEIGHT OF TENDER—40 tons 0 cwt. Full.
18 ,, 5 ,, Empty

"SAINT" CLASS.

No.	Name.	No.	Name.
2900	William Dean	2939	Croome Court
2901	Lady Superior	2940	Dorney Court
2902	Lady of the Lake	2941	Easton Court
2903	Lady of Lyons	2942	Fawley Court
2904	Lady Godiva	2943	Hampton Court
2905	Lady Macbeth	2944	Highnam Court
2906	Lady of Lynn	2945	Hillingdon Court
2907	Lady Disdain	2946	Langford Court
2908	Lady of Quality	2947	Madresfield Court
2909	Lady of Provence	2948	Stackpole Court
2910	Lady of Shalott	2949	Stanford Court
2911	Saint Agatha	2950	Taplow Court
2912	Saint Ambrose	2951	Tawstock Court
2913	Saint Andrew	2952	Twineham Court
2914	Saint Augustine	2953	Titley Court
2915	Saint Bartholomew	2954	Tockenham Court
2916	Saint Benedict	2955	Tortworth Court
2917	Saint Bernard	2971	Albion
2918	Saint Catherine	2972	The Abbot
2919	Saint Cuthbert	2973	Robins Bolitho
2920	Saint David	2974	Lord Barrymore
2921	Saint Dunstan	2975	Sir Ernest Palmer
2922	Saint Gabriel	2976	Winterstoke
2923	Saint George	2977	Robertson
2924	Saint Helena	2978	Kirkland
2926	Saint Nicholas	2979	Quentin Durward
2927	Saint Patrick	2980	Coeur de Lion
2928	Saint Sebastian	2981	Ivanhoe
2929	Saint Stephen	2982	Lalla Rookh
2930	Saint Vincent	2983	Redgauntlet
2931	Arlington Court	2984	Guy Mannering
2932	Ashton Court	2985	Peveril of the Peak
2933	Bibury Court	2986	Robin Hood
2934	Butleigh Court	2987	Bride of Lammermoor
2935	Caynham Court	2988	Rob Roy
2936	Cefntilla Court	2989	Talisman
2937	Clevedon Court	2990	Waverley
2938	Corsham Court	2998	Ernest Cunard

Serial Numbers :—
3800 — 3839.

3800 Class. Type 4—4—0.

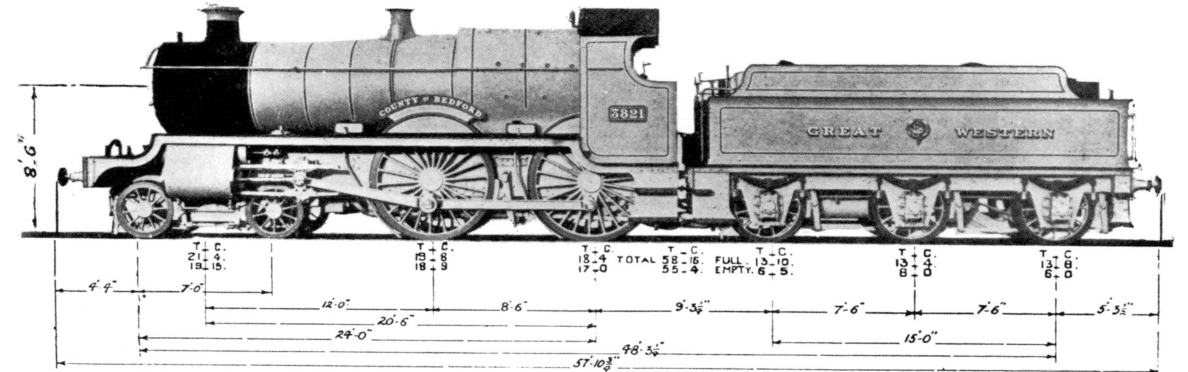

CYLINDERS—Diam., 18″ ; Stroke, 30″.
BOILER—Barrel, 11′0″ ; Diam. Outs., 5′ 6″ and 4′ $10\frac{3}{4}$″.
FIREBOX—Outs., 7′ 0″ by 5′ 9″ and 4′ 0″ ; Ins., 6′ $2\frac{1}{2}$″ by 4′ $8\frac{7}{8}$″ and 3′ $2\frac{5}{8}$″ ; Height, 6′ $6\frac{3}{8}$″ and 5′ $0\frac{3}{8}$″.
TUBES—Superheater Tubes, No. 84 Diam., 1″ ; Length, 11 $5\frac{3}{8}$″. Fire Tubes, No. 235 : Diam., $1\frac{5}{8}$″. No. 14 : Diam., $5\frac{1}{8}$″ ; Length 11′ 4 7/16″.

TOTAL WEIGHT OF ENGINE—58 tons 16 cwt. Full.
55 ,, 4 ,, Empty.

HEATING SURFACE—Superheater Tubes 191·79 sq. ft. Fire Tubes, 1,349·64 sq. ft. Firebox, 128·72 sq. ft. Total, 1670·15 sq. ft.
AREA OF FIREGRATE—20·56 sq. ft.
WHEELS—Bogie, 3′ 2″ ; Coupled, 6′ $8\frac{1}{2}$″.
WATER CAPACITY OF TENDER—3,500 gallons.
WORKING PRESSURE—200 lbs.
TRACTIVE EFFORT—20,530 lbs.

TOTAL WEIGHT OF TENDER—40 tons 0 cwt. Full
18 ,, 5 ,, Empty.

"COUNTY" CLASS.

No.	Name.	No.	Name.
3800	County of Middlesex	3820	County of Worcester
3801	County Carlow	3821	County of Bedford
3802	County Clare	3822	County of Brecon
3803	County Cork	3823	County of Carnarvon
3804	County Dublin	3824	County of Cornwall
3805	County Kerry	3825	County of Denbigh
3806	County Kildare	3826	County of Flint
3807	County Kilkenny	3827	County of Gloucester
3808	County Limerick	3828	County of Hereford
3809	County Wexford	3829	County of Merioneth
3810	County Wicklow	3830	County of Oxford
3811	County of Bucks	3831	County of Berks
3812	County of Cardigan	3832	County of Wilts
3813	County of Carmarthen	3833	County of Dorset
3814	County of Chester	3834	County of Somerset
3815	County of Hants	3835	County of Devon
3816	County of Leicester	3836	County of Warwick
3817	County of Monmouth	3837	County of Stafford
3818	County of Radnor	3838	County of Glamorgan
3819	County of Salop	3839	County of Pembroke

2925 Class. Type 4—6—0.

CYLINDERS—Diam. 18½"; Stroke, 30".
BOILER—Barrel, 14' 10"; Diam. Outs., 5' 6" and 4' 10 13/16".
FIREBOX—Outs., 9' 0" by 5' 9" and 4' 0"; Ins., 8' 2 7/16" by 4' 9" and 3' 2⅝"; Height, 6' 6⅜" and 5' 0⅜".
TUBES—Superheater Tubes, No. 84; Diam., 1"; Length, 15' 3⅜". Fire Tubes, No. 14: Diam. 5⅛". No. 176 Diam., 2"; Length, 15' 2 7/16".

TOTAL WEIGHT OF ENGINE—72 tons 10 cwt. Full.
66 ,, 10 ,, Empty.

HEATING SURFACE—Superheater Tubes, 262·62 sq. ft. Fire Tubes 1686·60 sq. ft. Firebox, 154·78 sq. ft. Total, 2104·0 sq. ft.
AREA OF FIREGRATE—27·07 sq. ft.
WHEELS—Bogie, 3' 2"; Coupled 6' 0".
WATER CAPACITY OF TENDER—3,500 gallons.
WORKING PRESSURE—225 lbs.
TRACTIVE EFFORT—27,275 lbs.

TOTAL WEIGHT OF TENDER—40 tons 0 cwt. Full.
18 ,, 5 ,, Empty.

" SAINT MARTIN " CLASS

No.	Name.
2925	Saint Martin

Type 4—4—2. De Glehn Compound.

CYLINDERS—High Pressure: Diam., 14 3/16″ Stroke, 25 3/16″. Low Pressure: Diam., 23⅝″; Stroke, 25 3/16″.
BOILER—Barrel, 14′ 10″; Diam. Outs. 4′ 10 13/16″ and 5′ 6″.
FIREBOX—Outs., 9′ 0″ by 5′ 9″ and 4′ 0″; Ins. 8 2 7/16″ by 4′ 9″ and 3′ 2⅝″; Height 6′ 6⅜″ and 5′ 0⅜″.
TUBES—Superheater Tubes, No. 84: Diam., 1″; Length, 15′ 3⅜″. Fire Tubes, No. 176 Diam., 2″; Length, 15′ 2 7/16″. Fire Tubes No. 14: Diam., 5⅛″; Length, 15′ 2 7/16″.

TOTAL WEIGHT OF ENGINE—70 tons 14 cwt. Full.
65 ,, 15 ,, Empty.

HEATING SURFACE—Superheater Tubes, 262·62 sq. ft. Fire Tubes, 1,686·60 sq. ft. Firebox, 154·78 sq. ft. Total 2,104·0 sq. ft.
AREA OF FIREGRATE—27·07 sq. ft.
WHEELS—Bogie, 3′ 2″; Coupled 6′ 8½″; Trailing 4′ 8⅜″.
WATER CAPACITY OF TENDER—3,500 gallons.
WORKING PRESSURE—225 lbs.
TRACTIVE EFFORT—26,935 lbs.

TOTAL WEIGHT OF TENDER—40 tons 0 cwt. Full.
18 ,, 5 ,, Empty.

" FRENCH " COMPOUND CLASS.

No	Name.
104	Alliance

GREAT WESTER

6000 ("KING") CLASS
4-6-0 (4 CYL.) TYPE. 6'-6" WHEELS.
TOTAL WEIGHT 135 - 14. – TRACTIVE EFFORT 40300 LBS.
ROUTE COLOUR – RED. –

4073 ("CASTLE") CLASS
4-6-0 (4 CYL.) TYPE. 6'-8½" WHEELS.
TOTAL WEIGHT 126 - 11 – TRACTIVE EFFORT 31625 LBS.
ROUTE COLOUR – RED. – GROUP LETTER "D".

2925 ("SAINT MARTIN") CLASS.
4-6-0 TYPE. 6'-0" WHEELS.
TOTAL WEIGHT 112 - 10. – TRACTIVE EFFORT 27275 LBS
ROUTE COLOUR –, RED. – GROUP LETTER "D".

4700 CLASS
2-8-0 TYPE. 5'-8" WHEELS.
TOTAL WEIGHT 122 - 0. – TRACTIVE EFFORT 30460 LBS.
ROUTE COLOUR – RED. – GROUP LETTER "D".

3800 ("COUNTY") CLASS
4-4-0 TYPE. 6'-8½" WHEELS.
TOTAL WEIGHT 98 - 16 – TRACTIVE EFFORT 20530 LBS.
ROUTE COLOUR – RED. – GROUP LETTER "C".

2221 CLASS
4-4-2T TYPE. 6'-8½ WHEELS.
TOTAL WT. 75 - 0.– TRACTIVE EFFORT 20530 LBS.
ROUTE COLOUR–RED. – GROUP LETTER "C".

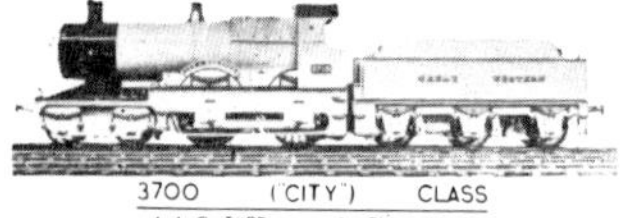

3700 ("CITY") CLASS
4-4-0 TYPE. 6'-8½ WHEELS
TOTAL WEIGHT 95 - 6. – TRACTIVE EFFORT 17790 LBS.
ROUTE COLOUR – RED. – GROUP LETTER "A".

4100 ("FLOWER") CLASS
4-4-0 TYPE. 6'-8½" WHEELS.
TOTAL WEIGHT 93 - 6. – TRACTIVE EFFORT 17790 LBS.
ROUTE COLOUR – RED. – GROUP LETTER "A".

2600 ("ABERDARE") CLASS
2-6-0 TYPE 4'-7½" WHEELS.
TOTAL WEIGHT 93 - 10 – TRACTIVE EFFORT 25800 LBS
ROUTE COLOUR – BLUE – GROUP LETTER "D"

2301 CLASS
0-6-0 TYPE 5'-2" WHEELS.
TOTAL WEIGHT 73 - 11 – TRACTIVE EFFORT 18140 LBS.
UNCOLOURED – GROUP LETTER "A"

TYPES OF LOCOMOTIV

AILWAY.

4000 ("STAR") CLASS
4-6-0 4 CYL. TYPE. 6'-8½" WHEELS.
TOTAL WEIGHT 115-12 - TRACTIVE EFFORT 27800 LBS.
ROUTE COLOUR - RED. - GROUP LETTER "D".

2900 ("SAINT") CLASS
4-6-0 TYPE. 6'-8½" WHEELS.
TOTAL WEIGHT 112-0. - TRACTIVE EFFORT 24395 LBS.
ROUTE COLOUR - RED. - GROUP LETTER "C".

4300 CLASS
2-6-0 TYPE. 5'-8" WHEELS.
TOTAL WEIGHT 102-0 - TRACTIVE EFFORT 25670 LBS.
ROUTE COLOUR - BLUE. - GROUP LETTER "D".

2800 CLASS
2-8-0 TYPE. 4'-7½" WHEELS.
TOTAL WEIGHT 115-10 - TRACTIVE EFFORT 35380 LBS.
ROUTE COLOUR - BLUE - GROUP LETTER "E".

3100 CLASS
2-6-2T TYPE. 5'-8" WHEELS.
L WT. 81-2. - TRACTIVE EFFORT 25670 LBS.
TE COLOUR-RED. - GROUP LETTER "D".

4200 CLASS
2-8-0T TYPE. 4'-7½" WHEELS.
TOTAL WT. 82-2. - TRACTIVE EFFORT 33170 LBS.
ROUTE COLOUR-RED. - GROUP LETTER "E".

4500 CLASS
2-6-2T TYPE. 4'-7½" WHEELS.
TOTAL WT. 61-0. TRACTIVE EFFORT 21250 LBS.
ROUTE COLOUR- YELLOW.- GROUP LETTER "C".

3300 & 3400 ("BULLDOG") CLASS
4-4-0 TYPE. 5'-8" WHEELS.
TOTAL WEIGHT 88-11. - TRACTIVE EFFORT 21060 LBS.
ROUTE COLOUR - BLUE. - GROUP LETTER "B".

3252 ("DUKE") CLASS
4-4-0 TYPE. 5'-8" WHEELS.
TOTAL WEIGHT 81-11. - TRACTIVE EFFORT 18955 LBS.
ROUTE COLOUR - YELLOW.- GROUP LETTER "B".

5600 CLASS
0-6-2T TYPE 4'-7½" WHEELS
WT. 68-12 - T.E. 25800 LBS.
COLOUR - RED - GROUP LETTER "D".

2700 CLASS
0-6-0T TYPE. 4'-7½" WHEELS.
TOTAL WT. 46-0 - T.E. 20260 LBS.
ROUTE COLOUR - BLUE.- GROUP LETTER "A".

1901 CLASS
0-6-0T TYPE. 4'-1½" WHEELS.
TOTAL WT. 35-18 - T.E. 17410 LBS.
UNCOLOURED. - UNGROUPED.

1361 CLASS
0-6-0T TYPE. 3'-8" WHEELS.
TOTAL WT. 35-4. - T.E. 14835 LBS.
UNCOLOURED. - UNGROUPED.

EPRODUCED TO SAME SCALE.

Serial Numbers :—
4000 — 4072.

4000 Class. Type 4—6—0.

CYLINDERS—Four: Diam., 15″; Stroke, 26″.
BOILER—Barrel 14′ 10″; Diam. Outs., 4′ 10 13/16″ and 5′ 6″.
FIREBOX—Outs., 9′ 0″ by 5′ 9″ and 4′ 0″; Ins., 8′ 2 7/16″ by 4′ 9″ and 3′ 2⅝″; Height, 6′ 6⅜″ and 5′ 0⅜″.
TUBES—Superheater Tubes, No. 84: Diam., 1″; Length, 15′ 3⅜″. Fire Tubes, No. 176: Diam., 2″; Length, 15′ 2 7/16″. Fire Tubes, No. 14 Diam., 5⅛″; Length 15′ 2 7/16″

TOTAL WEIGHT OF ENGINE—75 tons 12 cwt. Full.
70 ,, 3 ,, Empty.

HEATING SURFACE—Superheater Tubes, 262·62 sq. ft. Fire Tubes, 1,686·60 sq. ft. Firebox, 154·78 sq. ft. Total, 2,104·00 sq. ft.
AREA OF FIREGRATE—27·07 sq. ft.
WHEELS—Bogie 3′ 2″; Coupled, 6′ 8½″.
WATER CAPACITY OF TENDER—3,500 gallons.
WORKING PRESSURE—225 lbs.
TRACTIVE EFFORT—27,800 lbs.

TOTAL WEIGHT OF TENDER—40 tons 0 cwt. Full.
18 ,, 5 ,, Empty.

" STAR " CLASS

No.	Name.	No.	Name.
4000	North Star	4037*	Queen Philippa
4001	Dog Star	4038	Queen Berengaria
4002	Evening Star	4039	Queen Matilda
4003	Lode Star	4040	Queen Boadicea
4004	Morning Star	4041	Prince of Wales
4005	Polar Star	4042	Prince Albert
4006	Red Star	4043	Prince Henry
4007	Rising Star	4044	Prince George
4008	Royal Star	4045	Prince John
4009*	Shooting Star	4046	Princess Mary
4010	Western Star	4047	Princess Louise
4011	Knight of the Garter	4048	Princess Victoria
4012	Knight of the Thistle	4049	Princess Maud
4013	Knight of St. Patrick	4050	Princess Alice
4014	Knight of the Bath	4051	Princess Helena
4015	Knight of St. John	4052	Princess Beatrice
4016*	Knight of the Golden Fleece	4053	Princess Alexandra
4017	Knight of Liège	4054	Princess Charlotte
4018	Knight of the Grand Cross	4055	Princess Sophia
4019	Knight Templar	4056	Princess Margaret
4020	Knight Commander	4057	Princess Elizabeth
4021	British Monarch	4058	Princess Augusta
4022	Belgian Monarch	4059	Princess Patricia
4023	Danish Monarch	4060	Princess Eugenie
4024	Dutch Monarch	4061	Glastonbury Abbey
4025	Italian Monarch	4062	Malmesbury Abbey
4026	Japanese Monarch	4063	Bath Abbey
4027	Norwegian Monarch	4064	Reading Abbey
4028	Rumanian Monarch	4065	Evesham Abbey
4029	Spanish Monarch	4066	Malvern Abbey
4030	Swedish Monarch	4067	Tintern Abbey
4031	Queen Mary	4068	Llanthony Abbey
4032*	Queen Alexandra	4069	Westminster Abbey
4033	Queen Victoria	4070	Neath Abbey
4034	Queen Adelaide	4071	Cleeve Abbey
4035	Queen Charlotte	4072	Tresco Abbey
4036	Queen Elizabeth		

* Rebuilt as " Castle " Class.

Serial Numbers :—

4073 — 4099.
5000 — 5012.

4073 Class. Type 4—6—0.

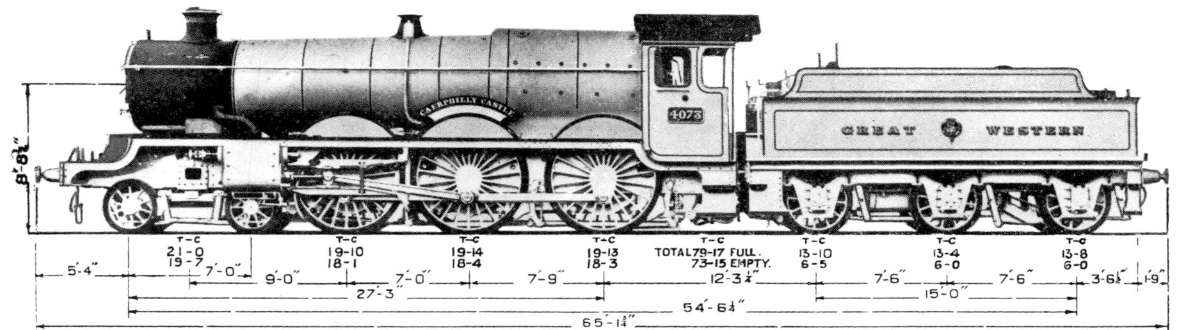

Cylinders—Four. Diam., 16"; Stroke 26".
Boiler—Barrel 14' 10"; Diam. Outs., 5' 9" and 5' 1 15/16".
Firebox—Outs., 10' 0" by 6' 0" and 4' 0"; Ins., 9' 2 7/16" by 5' 0⅛" and 3' 2⅝"; Height, 6' 8⅞" and 5' 3⅞".
Tubes—Superheater Tubes, No. 84: Diam., 1"; Length, 15' 3⅜". Fire Tubes, No. 201: Diam, 2"; No. 14: Diam., 5⅛"; Length, 15' 2 7/16".

Total Weight of Engine—79 tons 17 cwt. Full.
73 ,, 15 ,, Empty.

Heating Surface—Superheater Tubes, 262·62 sq. ft. Fire Tubes, 1,885·62 sq. ft. Firebox 163·76 sq. ft. Total, 2,312·0 sq. ft.
Area of Firegrate—30·28 sq. ft.
Wheels—Bogie, 3' 2"; Coupled, 6' 8½".
Water Capacity of Tender—3,500 gallons.
Working Pressure—225 lbs.
Tractive Effort—31,625 lbs.

Total Weight of Tender—40 tons 0 cwt. Full.
18 ,, 5 ,, Empty.

"CASTLE" CLASS.

No.	Name	No.	Name.
111	Viscount Churchill	4093	Dunster Castle
4073	Caerphilly Castle	4094	Dynevor Castle
4074	Caldicot Castle	4095	Harlech Castle
4075	Cardiff Castle	4096	Highclere Castle
4076	Carmarthen Castle	4097	Kenilworth Castle
4077	Chepstow Castle	4098	Kidwelly Castle
4078	Pembroke Castle	4099	Kilgerran Castle
4079	Pendennis Castle	5000	Launceston Castle
4080	Powderham Castle	5001	Llandovery Castle
4081	Warwick Castle	5002	Ludlow Castle
4082	Windsor Castle	5003	Lulworth Castle
4083	Abbotsbury Castle	5004	Llanstephan Castle
4084	Aberystwyth Castle	5005	Manorbier Castle
4085	Berkeley Castle	5006	Tregenna Castle
4086	Builth Castle	5007	Rougemont Castle
4087	Cardigan Castle	5008	Raglan Castle
4088	Dartmouth Castle	5009	Shrewsbury Castle
4089	Donnington Castle	5010	Restormel Castle
4090	Dorchester Castle	5011	Tintagel Castle
4091	Dudley Castle	5012	Berry Pomeroy Castle
4092	Dunraven Castle		

Four of the "Star" class have been rebuilt as "Castles"—see page 35.

G R

THIS ENGINE
No. 4082 "WINDSOR CASTLE" WAS BUILT AT
SWINDON IN APRIL 1924
AND WAS DRIVEN FROM THE WORKS TO THE STATION BY
HIS MAJESTY KING GEORGE V
ACCOMPANIED BY QUEEN MARY
ON THE OCCASION OF THE VISIT OF THEIR MAJESTIES
TO THE GREAT WESTERN RAILWAY WORKS AT
SWINDON ON APRIL 28TH 1924.

WITH THEIR MAJESTIES ON THE FOOTPLATE WERE

VISCOUNT CHURCHILL.	CHAIRMAN.
SIR FELIX POLE.	GENERAL MANAGER.
Mr. C.B.COLLETT.	CHIEF MECHANICAL ENGINEER.
LOCOMOTIVE INSPECTOR.	G.H.FLEWELLEN.
ENGINE DRIVER.	E.R.B.ROWE.
FIREMAN.	A.W.COOK.

THIS MEDAL is issued on the one hundredth birthday of the Baltimore & Ohio Railroad Company to commemorate not merely an important milestone in its own history but the rounding out of a century of a definite American railroad achievement.

The Baltimore & Ohio was the first American railroad to operate its line for the public handling of passengers and freight. This was early in 1830. In all the one hundred years of its life it has changed neither its corporate name, its charter nor its fundamental organization.

The obverse of the medal depicts one of the most modern trains of the Baltimore & Ohio—*The Capitol Limited* or *The National Limited*—drawn by one of the largest and most modern passenger locomotives built. The spirit of Transportation guides the locomotive in its onward flight.

The reverse shows the *Tom Thumb*, designed by Alderman Peter Cooper of New York and the first steam locomotive to be built in the United States, even though it was never put into practical service.

Mr. Hans Schuler, director of the Maryland Institute in Baltimore, is the sculptor who designed the medal and it was reproduced direct from his models by the Medallic Art Company of New York City.

Serial Numbers :—
6000 — 6019.

6000 Class. Type 4—6—0.

CYLINDERS—Four: Diam., 16¼"; Stroke 28".
BOILER—Barrel, Length 16' 0"; Diam. Outs., 6' 0" and 5' 6¼".
FIREBOX—Length Outs. 11' 6".

TOTAL WEIGHT OF ENGINE—89 tons 0 cwt. Full.
81 ,, 10 ,, Empty.

HEATING SURFACE—2,514 sq. ft. Total
AREA OF FIREGRATE—34·3 sq. ft.
WHEELS—Bogie, 3' 0"; Coupled 6' 6".
WATER CAPACITY OF TENDER—4,000 gallons.
WORKING PRESSURE—250 lbs.
TRACTIVE EFFORT—40,300 lbs.

TOTAL WEIGHT OF TENDER 46 tons 14 cwt. Full.
22 ,, 10 ,, Empty.

" KING " CLASS.

No.	Name.	No.	Name
6000	King George V	6010	King Charles I
6001	King Edward VII	6011	King James I
6002	King William IV	6012	King Edward VI
6003	King George IV	6013	King Henry VIII
6004	King George III	6014	King Henry VII
6005	King George II	6015	King Richard III
6006	King George I	6016	King Edward V
6007	King William III	6017	King Edward IV
6008	King James II	6018	King Henry VI
6009	King Charles II	6019	King Henry V

" King George V." with the Bell presented by the Baltimore and Ohio Railroad Company at the Centenary Celebration, September 24th—October 15th, 1927.

The Commemoration medal of the Baltimore and Ohio Exhibition fixed above the number plate of " King George V."

Serial Numbers :—

2301 — 2360.
2381 — 2580.

2301 Class. Type 0—6—0.

CYLINDERS—Diam., $17\frac{1}{2}$″; Stroke, 24″;
BOILER—Barrel, 10′ 3″; Diam. Outs. 4′ 5″ and 4′ 4″.
FIREBOX—Outs., 5 4″ by 4′ $7\frac{1}{8}$″ and 4′ 0″; Ins., 4′ 7 7/16″ by 3′ 8″ and 3′ 4″; Height, 6′ $0\frac{1}{2}$″.
TUBES—Superheater Tubes, No. 36: Diam., 1″; Length, 10′ $4\frac{3}{8}$″. Fire Tubes, No. 195: Diam., $1\frac{5}{8}$″; Length, 10′ 6 11/16″. Fire Tubes, No. 6 Diam., $5\frac{1}{8}$″; Length, 10′ 6 11/16″.

HEATING SURFACE—Superheater Tubes, 75·30 sq. ft. Fire Tubes, 960·85 sq. ft. Firebox, 106·45 sq. ft. Total, 1,142·60 sq. ft.
AREA OF FIREGRATE—15·45 sq. ft.
WHEELS—Coupled, 5′ 2″.
WATER CAPACITY OF TENDER—3,000 gallons.
WORKING PRESSURE—180 lbs.
TRACTIVE EFFORT—18 140 lbs.

2700 Class. Type $\frac{0—6—0.}{T}$

Serial Numbers :—
2700
2721 — 2799.

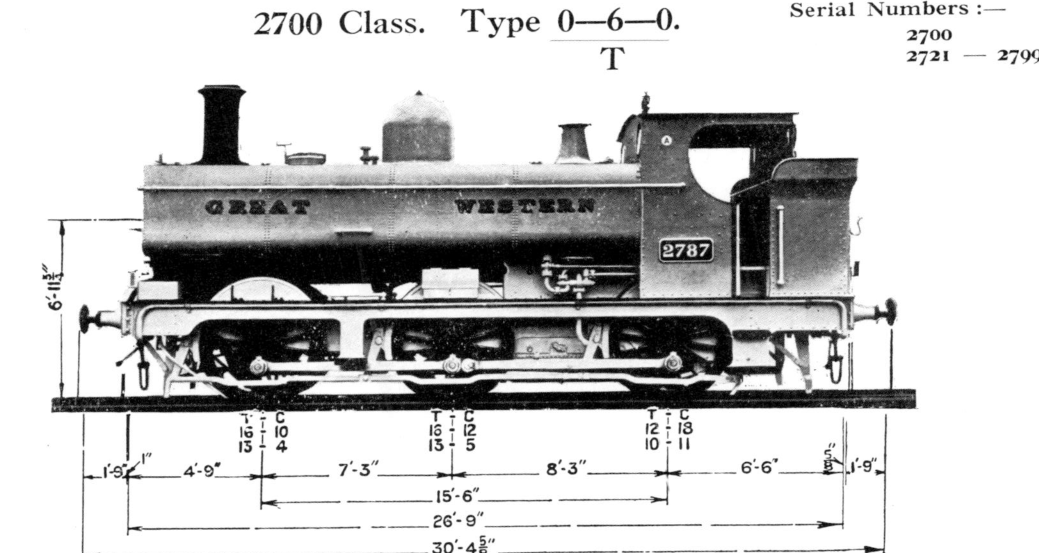

CYLINDERS—Diam., 17½″; Stroke, 24″.
BOILER—Barrel, 10′ 3″; Diam. Outs., 4′ 5″ and 4′ 4″.
FIREBOX—Outs., 5′ 4″ by 4′ 0″; Ins., 4′ 7 7/16″ by 3′ 4″; Height, 6′ 0½″.
TUBES—Superheater Tubes, No. 36: Diam., 1″; Length, 10. 4⅜″. Fire Tubes, No. 195: Diam., 1⅝″; Length, 10′ 6 11/16″. Fire Tubes, No. 6; Diam., 5⅛″; Length, 10′ 6 11/16″.

HEATING SURFACE—Superheater Tubes, 75·30 sq. ft. Fire Tubes 960·85 sq. ft. Firebox, 106·45 sq. ft Total, 1,142·60 sq ft.
AREA OF FIREGRATE—15·45 sq. ft.
WHEELS—Coupled, 4′ 7½″.
WATER CAPACITY OF TANK—1,200 gallons.
WORKING PRESSURE—180 lbs.
TRACTIVE FORCE—20,260 lbs.

TOTAL WEIGHT OF ENGINE—46 tons 0 cwt. Full
37 „ 0 „ Empty.

Serial Numbers :—
2600 — 2680.

2600 Class. Type 2—6—0.

Cylinders—Diam., 18″; Stroke 26″.
Boiler Barrel 11′ 0″; Diam. Outs., 4′ $10\frac{3}{4}$″ and 5′ 6″.
Firebox—Outs., 7′ 0″ by 5′ 9″ and 4′ 0″; Ins., 6′ $2\frac{1}{2}$″ by 4′ $8\frac{7}{8}$″ and 3′ $2\frac{5}{8}$″; Height, 6′ $6\frac{3}{8}$″ and 5′ $0\frac{3}{8}$″.
Tubes—Superheater Tubes, No. 84: Diam., 1″; Length, 11′ $5\frac{3}{8}$″. Fire Tubes, No. 235: Diam., $1\frac{5}{8}$″; Length, 11 4 7/16″. Fire Tubes, No. 14: Diam. $5\frac{1}{8}$″; Length, 11′ 4 7/16″.

Total Weight of Engine—56 tons 15 cwt. Full.
51 ,, 5 ,, Empty.

Heating Surface—Superheater Tubes, 191·79 sq. ft. Fire Tubes 1,349·64 sq. ft. Firebox, 128·72 sq. ft. Total, 1,670·15 sq. ft.
Area of Firegrate—20·56 sq. ft.
Wheels—Pony, 2′ 8″; Coupled, 4′ $7\frac{1}{2}$″.
Water Capacity of Tender—3,000 gallons.
Working Pressure—200 lbs.
Tractive Effort—25,800 lbs.

Total Weight of Tender—36 tons 15 cwt. Full.
17 ,, 9 ,, Empty.

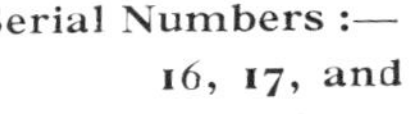

Type 0—6—4 T

Serial Numbers :—
16, 17, and 18.

Great Western Railway.

Named Engines.
0–6–4 Type.

SUNDRY
(Not illustrated)

No.	Name.	No.	Name.
16	Hercules	1308	Lady Margaret
17	Cyclops	1473	Fair Rosamond
18	Steropes	1813	Holmwood

CYLINDERS—Diam., 16″; Stroke, 24″.
BOILER—BARREL, 10′ 0″; Diam. Outs. 3′ 10″ and 3′ 9½″.
FIREBOX—Outs., 4′ 0″ by 4′ 0″; Ins., 3′ 3 15/16″ by 3′ 4″; Height, 5′ 5½″
TUBES—No. 207: Diam., 1⅝″: Length, 10′ 3 3/16″.
HEATING SURFACE—Tubes, 904·47 sq. ft. Firebox, 76·28 sq. ft. Total, 980·75 sq. ft.
AREA OF FIREGRATE—11·16 sq. ft.
WHEELS—Coupled, 4′ 1½″; Bogie, 2′ 8″.
WATER CAPACITY OF TANK—1,200 gallons.
WORKING PRESSURE—165 lbs.
TRACTIVE EFFORT—17,410 lbs.
LOAD—6 tons at 18′•0″ radius (double chain); 9 tons at 12′ 0″ radius (treble chain).

TOTAL WEIGHT OF ENGINE—63 tons 12 cwt. Full.
55 ,, 17 ,, Empty.

Serial Numbers :—
2800 — 2883.

2800 Class. Type 2—8—0.

CYLINDERS—Diam., 18½″ ; Stroke, 30″.
BOILER—Barrel, 14′ 10″ ; Diam. Outs., 5′ 6″ and 4′ 10 13/16″.
FIREBOX—Outs., 9′ 0″ by 5′ 9″ and 4′ 0″ ; Ins., 8′ 2 7/16″ by 4′ 9″ and 3′ 2⅝″ ; Height, 6′ 6⅜″ by 5′ 0⅜″.
TUBES—Superheater Tubes, No. 84 : Diam., 1″ ; Length, 15′ 3⅜″. Fire Tubes, No. 176 : Diam., 2″ : Length, 15′ 2 7/16″. Fire Tubes, No. 14 : Diam., 5⅛″ : Length, 15′ 2 7/16″.

TOTAL WEIGHT OF ENGINE—75 tons 10 cwt. Full.
70 ,, 2 ,, Empty.

HEATING SURFACE—Superheater Tubes, 262·62 sq. ft. Fire Tubes, 1,686·60 sq. ft. Firebox 154·78 sq. ft. Total, 2104·0 sq. ft.
AREA OF FIREGRATE—27·07 sq. ft.
WHEELS—Pony, 3′ 2″ ; Coupled, 4′ 7½″.
WATER CAPACITY OF TENDER—3,500 gallons.
WORKING PRESSURE—225 lbs.
TRACTIVE EFFORT—35,380 lbs.

TOTAL WEIGHT OF TENDER—40 tons 0 cwt. Full.
18 ,, 5 ,, Empty.

Note:—Some with modified weight distribution re-numbered 51xx.

Serial Numbers: 3150 — 3190.

3150 Class. Type $\frac{2—6—2.}{T}$

Cylinders—Diam., $18\frac{1}{2}''$; Stroke, 30″.
Boiler—Barrel, 11′ 0″; Diam. Outs. 5′ 6″ and 4′ $10\frac{3}{4}''$.
Firebox—Outs., 7′ 0″ by 5′ 9″ and 4′ 0″; Ins., 6′ $2\frac{1}{2}''$ by 4′ $8\frac{7}{8}''$ and 3′ $2\frac{5}{8}''$; Height, 6′ $6\frac{3}{8}''$ and 5′ $0\frac{3}{8}''$.
Tubes—Superheater Tubes, No. 84: Diam., 1″: Length, 11′ $5\frac{3}{8}''$. Fire Tubes, No 235: Diam., $1\frac{5}{8}''$; Length, 11′ 4 7/16″. Fire Tubes, No. 14: Diam. $5\frac{1}{8}''$; Length, 11 4 7/16″.

Heating Surface—Superheater Tubes, 191·79 sq. ft. Fire Tubes, 1,349·64 sq. ft. Firebox 128·72 sq. ft. Total, 1,670·15 sq. ft.
Area of Firegrate—20·56 sq. ft.
Wheels—Pony Truck, 3′ 2″; Coupled, 5′ 8″; Radial Truck, 3′ 8″.
Water Capacity of Tanks—2,000 gallons.
Working Pressure—200 lbs.
Tractive Effort—25,670 lbs.

Total Weight of Engine—81 tons 12 cwt. Full.
64 ,, 13 ,, Empty.

Serial Numbers :—
2221 — 2250.

2200 Class. Type 4—4—2. T

CYLINDERS—Diam. 18″; Stroke, 30″.
BOILER—Barrel 11′ 0″; Diam. Outs., 5′ 0½″ and 4 5⅛″.
FIREBOX—Outs., 7′ 0″ by 5′ 3″ and 4′ 0″; Ins., 6′ 2 11/16″ by 4′ 3½″ and 3′ 2¾″; Height 6′ 0 7/16″ and 5′ 0 7/16″.
TUBES—Superheater Tubes, No. 36: Diam., 1″; Length 11′ 5⅜″; Fire Tubes, No. 218: Diam., 1⅝″; Length, 11′ 4 5/16″. Fire Tubes, No. 6: Diam., 5⅛″; Length, 11′ 4 5/16″.

HEATING SURFACE—Superheater Tubes, 82·20 sq. ft. Fire Tubes, 1,144·95 sq. ft. Firebox, 121·80 sq. ft. Total, 1,348·95 sq. ft.
AREA OF FIREGRATE—20·35 sq. ft.
WHEELS—Bogie, 3′ 2″; Coupled, 6′ 8½″; Radial Truck, 3′ 8″.
WATER CAPACITY OF TANK—2,000 gallons.
WORKING PRESSURE—200 lbs.
TRACTIVE EFFORT—20,530 lbs.

TOTAL WEIGHT OF ENGINE—75 tons 0 cwt. Full.
59 ,, 10 ,, Empty.

4500 Class. Type $\frac{2—6—2.}{T}$

Serial Numbers :—
4500 — 4599.
5500 — 5524.

CYLINDERS—Diam., 17″; Stroke, 24″.
BOILER—Barrel, 10′ 6″; Diam. Outs., 4′ 9½″ and 4′ 2″.
FIREBOX—Outs., 5′ 10″ by 5′ 0″ and 4′ 0″; Ins., 5′ 0 11/16″ by 4′ 0″ and 3′ 2¾″; Height, 5′ 8″ and 4′ 8″.
TUBES—Superheater Tubes, No. 36; Diam., 1″; Length, 10′ 10⅛″. Fire Tubes, No. 196: Diam., 1⅝″; Length, 10′ 10 5/16″. Fire Tubes, No. 6: Diam., 5⅛″; Length 10′ 10 5/16″.

HEATING SURFACE—Superheater Tubes, 77·64 sq. ft. Fire Tubes 992·51 sq. ft. Firebox, 94·25 sq. ft. Total 1,164·40 sq. ft.
AREA OF FIREGRATE—16·6 sq. ft.
WHEELS—Pony Truck, 3′ 2″; Coupled, 4′ 7½″; Pony Truck, 3′ 2″.
WATER CAPACITY OF TANK—1,300 gallons.
WORKING PRESSURE—200 lbs.
TRACTIVE EFFORT—21,250 lbs.

TOTAL WEIGHT OF ENGINE—61 tons 0 cwt. Full.
49 ,, 15 ,, Empty.

Type 0—6—0. / T

Serial Numbers :
1361 — 1365.

CYLINDERS—Diam., 16″; Stroke, 20″.
BOILER—Barrel, 8′ 2″; Diam. Outs., 3′ 9⅛″ and 3′ 10″.
FIREBOX—Outs., 3′ 11″ by 4′ 0″; Ins., 3′ 3 1/16″ by 3′ 4⅛″; Height, 5′ 3½″.
TUBES—No. 207: Diam., 1⅝″; Length, 8′ 5⅛″.

HEATING SURFACE—Tubes, 815·5 sq. ft. Firebox, 74·75 sq. ft. Total, 890·25 sq. ft.
AREA OF FIREGRATE—10·71 sq. ft.
WHEELS—Leading, 3′ 8″; Driving, 3′ 8″; Trailing, 3′ 8″.
WATER CAPACITY OF TANK—800 gallons.
WORKING PRESSURE—150 lbs.
TRACTIVE EFFORT—14,835 lbs.

TOTAL WEIGHT OF ENGINE—35 tons 4 cwt. Full.
27 ,, 11 ,, Empty.

Note :—Some with modified weight distribution re-numbered 83xx.

Serial Numbers :—

4300 — 4399.
5300 — 5399.
6300 — 6399.
7300 — 7321.

4300 Class. Type 2—6—0.

CYLINDERS—Diam., $18\frac{1}{2}''$; Stroke, 30″.
BOILER—Barrel, 11′ 0″; Diam. Outs., 5′ 6″ and 4′ $10\frac{3}{4}''$.
FIREBOX—Outs., 7′ 0″ by 5′ 9″ and 4′ 0″ Ins., 6′ $2\frac{1}{2}''$ by 4′ $8\frac{7}{8}''$ and 3′ $2\frac{5}{8}''$; Height, 6′ $6\frac{3}{8}''$ and 5′ $0\frac{3}{8}''$.
TUBES—Superheater Tubes, No. 84; Diam., 1″; Length, 11′ $5\frac{3}{8}''$. Fire Tubes, No. 235: Diam., $1\frac{5}{8}''$; Length, 11′ 4 7/16′. Fire Tubes, No. 14: Diam., $5\frac{1}{8}''$; Length, 11′ 4 7/16″.

TOTAL WEIGHT OF ENGINE—62 tons 0 cwt. Full.
57 ,, 14 ,, Empty.

HEATING SURFACE—Superheater Tubes, 191·79 sq. ft. Fire Tubes, 1,349·64 sq. ft. Firebox, 128·72 sq. ft. Total, 1,670·15 sq. ft.
AREA OF FIREGRATE—20·56 sq. ft.
WHEELS—Pony, 3′ 2″; Coupled, 5′ 8″.
WATER CAPACITY OF TENDER—3,500 gallons.
WORKING PRESSURE—200 lbs.
TRACTIVE EFFORT—25,670 lbs.

TOTAL WEIGHT OF TENDER—40 tons 0 cwt. Full.
18 ,, 5 ,, Empty.

4200 Class. Type 2—8—0. T

Serial Numbers :—

4200 — 4299.

5200 — 5274.

CYLINDERS—Diam., 19″; Stroke, 30″.
BOILER—Barrel, 11′ 0″; Diam Outs., 5′ 6″ and 4′ 10$\frac{3}{4}$″
FIREBOX—Outs., 7′ 0″ by 5′ 9″ and 4′ 0″; Ins., 6′ 2$\frac{1}{2}$″ by 4′ 8$\frac{7}{8}$″ and 3′ 2$\frac{5}{8}$″; Height, 6′ 6$\frac{3}{8}$″ and 5′ 0$\frac{3}{8}$.
TUBES—Superheater Tubes, No. 84: Diam., 1″; Length, 11′ 5$\frac{3}{8}$″. Fire Tubes, No. 235: Diam., 1$\frac{5}{8}$″; Length, 11′ 4 7/16″. Fire Tubes, No. 14 Diam., 5$\frac{1}{8}$″, Length, 11′ 4 7/16″.

HEATING SURFACE—Superheater Tubes, 191·79 sq. ft. Fire Tubes, 1,349·64 sq. ft. Firebox, 128·72 sq. ft. Total, 1,670·15 sq. ft.
AREA OF FIREGRATE—20·56 sq. ft.
WHEELS—Pony Truck, 3′ 2″; Coupled, 4′ 7$\frac{1}{2}$″.
WATER CAPACITY OF TANK—1,800 gallons.
WORKING PRESSURE—200 lbs.
TRACTIVE EFFORT—33,170 lbs.

Serial Numbers :—
4700 — 4708.

4700 Class. Type 2—8—0.

CYLINDERS—Diam., 19″; Stroke, 30″.
BOILER—Barrel, 14′ 10″; Diam Outs, 6′ 0″ and 5′ 6″.
FIREBOX—Outs., 10′ 0″ by 6′ 3″ and 4′ 0″; Ins., 9′ 2⅜″ by 5′ 2⅜″ and 3′ 2⅝″; Height, 6′ 10⅜″ and 5′ 2⅜″.
TUBES—Superheater Tubes, No. 96: Diam, 1″; Length, 15′ 3⅜″; Fire Tubes, No. 218; Diam., 2″; No. 16; Diam., 5⅛″; Length, 15′ 2½″.

TOTAL WEIGHT OF ENGINE—82 tons 0 cwt. Full.
75 ,, 2 ,, Empty.

HEATING SURFACE—Superheater Tubes, 289·60 sq. ft. Fire Tubes 2,062·35 sq. ft. Firebox, 169·75 sq. ft. Total, 2,521·70 sq. ft.
AREA OF FIREGRATE—30·28 sq. ft.
WHEELS—Pony, 3′ 2″; Coupled, 5′ 8″.
WATER CAPACITY OF TENDER—3,500 gallons.
WORKING PRESSURE—225 lbs.
TRACTIVE EFFORT—30,460 lbs.

TOTAL WEIGHT OF TENDER 40 tons 0 cwt. Full.
18 ,, 5 ,, Empty.

5600 Class. Type 0—6—2. T

Serial Numbers :—

5600 — 5699.
6600 — 6629.

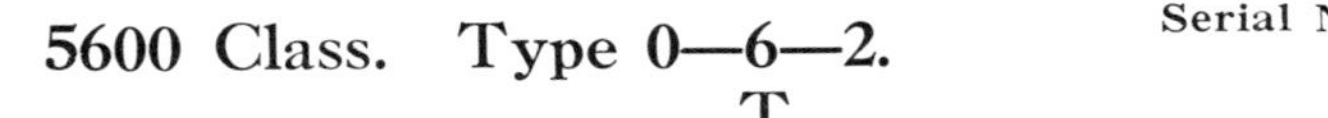

CYLINDERS—Diam., 18″; Stroke 26″.
BOILER—Barrel, 11′ 0″; Diam. Outs., 5′ 0½″ and 4′ 5⅛″.
FIREBOX—Outs., 7′ 0″ by 5′ 3″ and 4′ 0″ Ins., 6′ 2 11/16″ by 4′ 3½″ and 3′ 2¾″; Height, 6′ 0 7/16″ and 5′ 0 7/16″.
TUBES—Superheater Tubes, No. 36: Diam., 1″; Length, 11″ 5⅜″ Fire Tubes, No. 6: Diam., 5⅛; No. 218: Diam., 1⅝: Length, 11′ 4 5/16″

HEATING SURFACE—Superheater Tubes 82·20 sq. ft. Fire Tubes, 1,144·95 sq. ft. Firebox, 121·80 sq. ft. Total, 1,348·95 sq. ft.
AREA OF FIREGRATE—20·35 sq. ft.
WHEELS—Coupled, 4′ 7½″, Radial, 3′ 8″.
WATER CAPACITY OF TANKS—1,900 gallons.
WORKING PRESSURE—200 lbs.
TRACTIVE EFFORT—25,800 lbs.

TOTAL WEIGHT OF ENGINE—68 tons 12 cwt. Full.
53 ,, 12 ,, Empty.

ALPHABETICAL INDEX TO NAMED ENGINES.

Alphabetical Index to Named Engines—*continued.*

Alphabetical Index to Named Engines—*continued.*

Name.	No.	Page	Name.	No.	Page
Hercules	16	43	King James II ..	6008	41
Highclere Castle ..	4096	37	King Richard III ..	6015	41
Highnam Court ..	2944	25	King William III ..	6007	41
Hillingdon Court ..	2945	25	King William IV ..	6002	41
Hobart	3703	23	Kingfisher	3448	19
Holmwood	1813	43	Kingsbridge	3347	19
Hotspur	4108	21	Kirkland	2978	25
Hyacinth	4161	21	Kitchener	4124	21
Ilfracombe	3383	19	Knight Commander ..	4020	35
Inchcape	3430	19	Knight of Liège ..	4017	35
Isle of Guernsey ..	3312	19	Knight of St. John ..	4015	35
Isle of Jersey	3284	17	Knight of St. Patrick	4013	35
Isle of Tresco.. ..	3277	17	Knight of the Bath ..	4014	35
Italian Monarch ..	4025	35	Knight of the Garter	4011	35
Ivanhoe	2981	25	Knight of the Golden Fleece	4016	35
Jackdaw	3447	19			
Jamaica	3402	19	Knight of the Grand Cross	4018	35
James Mason	3413	19			
Japanese Monarch ..	4026	35	Knight of the Thistle	4012	35
John G. Griffiths ..	3412	19	Knight Templar ..	4019	35
John W. Wilson ..	3416	19	Lady Disdain ..	2907	25
Joseph Shaw	3431	19	Lady Godiva	2904	25
Jupiter..	3313	19	Lady Macbeth ..	2905	25
Katerfelto	3285	17	Lady Margaret ..	1308	43
Kekewich	4129	21	Lady of Lynn ..	2906	25
Kenilworth	3325	19	Lady of Lyons ..	2903	25
Kenilworth Castle ..	4097	37	Lady of Provence ..	2909	25
Kidwelly Castle ..	4098	37	Lady of Quality ..	2908	25
Kilgerran Castle ..	4099	37	Lady of Shalott ..	2910	25
Killarney	3708	23	Lady of the Lake ..	2902	25
King Charles I ..	6010	41	Lady Superior ..	2901	25
King Charles II ..	6009	41	Ladysmith	4127	21
King Edward IV ..	6017	41	Laira	3326	19
King Edward V ..	6016	41	Lalla Rookh	2982	25
King Edward VI ..	6012	41	Langford Court ..	2946	25
King Edward VII ..	6001	41	Launceston	3348	19
King George I ..	6006	41	Launceston Castle ..	5000	37
King George II ..	6005	41	Llandovery Castle ..	5001	37
King George III ..	6004	41	Llanstephan Castle ..	5004	37
King George IV ..	6003	41	Llanthony Abbey ..	4068	35
King George V ..	6000	41	Lobelia	4157	21
King Henry V ..	6019	41	Lode Star	4003	35
King Henry VI ..	6018	41	Lord Barrymore ..	2974	25
King Henry VII ..	6014	41	Lord Mildmay of Flete	3417	19
King Henry VIII ..	6013	41	Ludlow Castle ..	5002	37
King James I ..	6011	41	Lulworth Castle ..	5003	37

Alphabetical Index to Named Engines—*continued.*

Alphabetical Index to Named Engines—*continued.*

Alphabetical Index to Named Engines—*continued.*

Name.	No.	Page	Name.	No.	Page
Tor Bay	3279	17	Vancouver	3401	19
Toronto	3397	19	Viscount Churchill ..	111	37
Torquay	3360	19	Vulcan..	3318	19
Tortworth Court ..	2955	25	Walter Long	3374	19
Trefusis	3278	17	Warwick Castle ..	4081	37
Tregeagle	3359	19	Waverley	2990	25
Tregenna	3280	17	Western Star	4010	35
Tregenna Castle ..	5006	37	Westminster Abbey ..	4069	35
Tregothnan	3335	19	Weston-Super Mare ..	3439	19
Trelawney	3357	19	Weymouth	3319	19
Tremayne	3358	19	White	4138	21
Tre Pol and Pen ..	3265	17	William Dean.. ..	2900	25
Tresco Abbey ..	4072	35	Windsor Castle ..	4082	37
Trevithick	3264	17	Winnipeg	3400	19
Trinidad	3403	19	Winterstoke	2976	25
Twineham Court ..	2952	25	Wolseley	4137	21

MORE ABOUT ENGINES.

If this Volume has interested you, you cannot fail to be enthralled by the following three books, which have been specially written

FOR BOYS OF ALL AGES.

THE 10.30 LIMITED.

A Railway Book, 144 pages, 125 pictures.

Boys and girls everywhere recognize the interest, charm and wonder of the transit ways of steel, steam and speed. Also they demand information. For their enlightenment and edification this book has been produced in which the author assumes that he and his young reader are together making the non-stop railway run (Paddington to Plymouth) on the Cornish Riviera Express, familiarly known as "The 10.30 Limited." During this fascinating journey of the "10.30 Limited" the hows, whys and wherefores of railway working are pleasantly narrated and pictorially explained.

CAERPHILLY CASTLE.

A book of Railway Locomotives, 208 pages, 150 pictures.

This book, a companion volume to "The 10.30 Limited," has been written to meet a demand for more information about the railway locomotive. The present edition is almost exhausted, and to take its place

THE "KING" OF RAILWAY LOCOMOTIVES.

is being prepared. It traces the evolution of the Railway Locomotive from its earliest days to the production of the King George V. the most powerful passenger locomotive. Not only are the various component parts of the engine described and their functions explained,but the reader is taken for an imaginary tour of the works where the engine is built. For 1/- you will be enabled to answer most of the queries which youngsters always have regarding Railways. The book will be ready at the beginning of October, 1928.

Each **1/-** net.

Crown Octavo

ITS A PUZZLE

TO FIND SIMILAR VALUE AT ANYTHING APPROACHING THE PRICE.

YOU CAN'T DO IT.

BUT YOU CAN DO THE FOLLOWING JIG-SAW PUZZLES :

" KING GEORGE V."

A jig-saw puzzle which, when completed, forms a picture in colours of the super-locomotive " King George V." the most powerful passenger train locomotive in the British Isles.

The puzzle makes a direct appeal to the many thousands of engine-admirers.

(Copies of this puzzle, cut into 300 pieces to make it a real "teazer," are also available, price 3/6.)

" BRITAIN'S MIGHTIEST."

A picture which conveys something of the majesty of a Great Western Railway locomotive.

" CATHEDRAL."

Shewing on one side " Exeter Cathedral " from the picture by Fred Taylor, R.I. and on the other a railway map of England and Wales.

" SWANSEA DOCKS."

A unique bird's-eye view of a busy dock. The detail makes this a most interesting picture.

" OXFORD."

A part of this beautiful city of learning portrayed by a master of architectural drawing—Fred Taylor, R.I.

" ST. JULIEN."

A reproduction of the G.W.R.'s popular Channel Islands Service Steamship " St. Julien."

" SPEED."

Shewing a typical Great Western Express train travelling at speed.

" THE CORNISH RIVIERA EXPRESS."

Reproduction of the most famous passenger train in the world.

" THE FREIGHT TRAIN."

A representative Great Western Goods Express.

Price 2/6 each.

Pages from the 1938 edition

NOTE There is surprisingly little difference between the 1938 and 1946 editions and reprinting both would be pointless. There are of course certain locomotives listed in one and not the other; the *Bulldog* class had been substantially reduced in size by 1946, while the *County* class was new. But certain editorial contents dropped from the 1946 edition seem worth preserving and are therefore included with their original page numbers in the following pages.

FOREWORD

DURING the hundred years that the Great Western Railway has been in existence, there have apparently always been railway enthusiasts (young and old) keenly interested in G.W.R. locomotives in general and their nomenclature in particular. The hobby of observing and recording names of engines seems to have been adopted by large numbers of railway devotees quite half-a-century before the first issue of any publication on the subject.

As evidence of this, the pocket-book of a fourteen-years-old compiler, dated 1861, and headed "Names of Engines on the Great Western I have seen," recently came under notice. It included such names as *North Star, Firefly, Meridian* and *Eclipse* and, among many others, *Great Western,* the first engine completely built at Swindon Works (in 1846). A most remarkable fact about this unique record of seventy-seven years ago is that it contained the names of all Gooch's "eight-foot singles" that were in traffic at that time.

In those early days, G.W.R. locomotive enthusiasts made their own lists of engines from personal observation, and the following from an "old-time" engine-name collector and correspondent of the *Great Western Railway Magazine* throws some light upon the hobby as then pursued.

"To the small, exclusive, band of us who were devotees of this cult in those days, the art of engine-name collecting was no milk-and-water affair. To the enthusiasts it was a consuming fire and a source of continual enthralment ; and until adolescence gradually weaned us into other and more serious interests, some of us were almost in the fanatic stage. The cult demanded long and steady vigils, hazardous excursions, and often daring courage; it had withal, though not registered, nor even written, a complete complement of rules and regulations, to which we rigidly adhered, or knew ourselves to be dishonourable if we did not.

"We each possessed a pocket-book in which were inscribed the names in the respective order of 'capture.' We also essayed hand-sketched illustrations of types, but how far these conformed to the originals was always open to debate."

This account of G.W.R. engine-name collecting is more or less typical of many received, some from correspondents in the most distant parts of the world, testifying that the writers, though now belonging to the "old brigade," were once ardent and enthusiastic members of the G.W.R. engine-name collecting brotherhood, and have maintained their devotion to the Great Western Railway ever since.

Although the interest evinced in G.W.R. locomotive names was recognised from time to time by the inclusion of the names of new engines produced at Swindon Works in the pages of the *Great Western Railway Magazine,* no separate official list was published until 1911, when a slight volume of twenty-four pages (twelve of which were devoted to illustrations) appeared, entitled "Names of Engines." It contained the names and numbers of 461 named engines then in Great Western Railway service.

It speaks volumes for the interest taken in the subject that up to 1935 no fewer than thirteen issues of the G.W.R. Engine Book have been called for. In the later editions, the information as to names and numbers of engines was supplemented by giving their salient dimensions, etc. The provision of this additional matter at once gave the book a wider appeal, for it found a welcome among many who, though not primarily interested in locomotive nomenclature, were glad to be furnished with measurements and other particulars of the various types of standard G.W.R. locomotives.

Such has been the progress in locomotive construction on the Great Western Railway since the last edition of the book (1935) appeared that over a hundred-and-fifty deletions, alterations, and additions have been necessary in the list of named engines alone. These amendments have been embodied in the present edition and the opportunity has been taken to re-arrange and revise the text and to add some new features which, it is hoped, will be appreciated by all interested in G.W.R. engines.

W. G. C.

PASSENGER AND FREIGHT TRAIN NAMES

WHILST not strictly within the purview of a book entitled "G.W.R. Engines," the subject of train names is closely associated and is, perhaps, permissible of inclusion.

It may be that some G.W.R. trains were known by specific names before the coming of the famous "Flying Dutchman" (1862), but if that is the case the names do not appear to have survived. "Flying Dutchman" (11.45 a.m. Paddington to Exeter) was a history maker, for, when introduced, it travelled at the highest speed then known for any form of transport. It was later accelerated so as to arrive at Plymouth (via Bristol) $6\frac{1}{4}$ hours after leaving Paddington —a really wonderful timing for those days.

"Zulu" (3.0 p.m. from Paddington) was put on in 1879 and travelled at the same pace as "Flying Dutchman." These two were the G.W.R. crack trains up to the beginning of the twentieth century.

To-day, the most famous of the named passenger trains are :—

"Cornish Riviera Limited" (Described elsewhere)
"Cheltenham Flyer" (Described elsewhere)
"The Bristolian" (10.0 a.m. from Paddington) and
"Torbay Express" (12.0 noon from Paddington)

All these trains have fast timings, and perhaps it should be remarked that "The Bristolian" was introduced on September 9th, 1935, to commemorate the Centenary of the Great Western Railway and Bristol's association therewith. "The Bristolian" makes the double journey of 236 miles at average speeds of 67.6 and 67.1 miles an hour for the outward and return trips respectively.

These train names are self-explanatory, although perhaps a word or two may be said about "Cheltenham Flyer" which seems to have been christened spontaneously by the press and public. The train is officially known as "Cheltenham Spa Express," and it is on its "up" journey (from Cheltenham to

The " Bristolian " on Wharncliffe Viaduct—Locomotive " King Henry VIII."

Paddington) that it makes the record run of over $77\frac{1}{4}$ miles from Swindon (the last intermediate stop) to Paddington at an average speed of 71.3 miles an hour.

So much for the principal named G.W.R. passenger trains, but freight trains also have their names, and these may be described as unofficial christenings (or nick-names) bestowed by the staff associated with them. In many cases they are traditional, having been handed on to enginemen, guards, and shunters, by their predecessors who have long retired from active service.

A variety of features have contributed to the choice of names, but it is clear that the two principal considerations have been the territory served and the class of traffic carried. Many of the names are romantic, e.g., "The Flying Pig," "The Northern Flash," "The Drake" ; some are homely (and none the less effective), in which category "The Bacca," "The Mopper Up" and "The Spud" are notable.

Other names which more or less suggest the nature of the traffic carried or the starting point are "The Western Docker" (Bristol to Wolverhampton), "The Cocoa" (Bristol to Paddington), "The Cotswold" (Gloucester to Paddington), "The Carpet" (Kidderminster to Paddington), "The Biscuit" (Reading to Laira), and the "Up Jersey" (Weymouth to Paddington).

" Cheltenham Flyer."

CHELTENHAM FLYER

THE G.W.R. train which is known as "Cheltenham Flyer" has long been famous for its high speed, and its story is one of steady acceleration. Before the war, it ran "non-stop" from Kemble Junction to Paddington—91 miles—in 103 minutes. When the war was over and normal services restored, a stop was put in at Swindon and 85 minutes allowed for the 77¼ miles' run from there to Paddington. From July, 1923, the timing for the Swindon-Paddington run was reduced to 75 minutes or at an average speed of 61.8 miles per hour. At that speed it was the fastest "start to stop" run in the British Isles, and when in July, 1929, it was accelerated up to an average speed of 66.2 miles per hour—70 minutes being then allowed for doing the 77¼ miles—it took precedence of any other train in the world.

Two years later (1931) the train was again accelerated to do the Swindon-Paddington trip at an average speed of 69.2 miles an hour, and in September, 1932, it was still further speeded up, and the daily run from Swindon to Paddington was scheduled to take 65 minutes, giving the extraordinary average speed of 71.3 miles an hour for the whole trip of 77¼ miles. Incidentally, that was the first occasion in the history of railways that any train had been timed at over seventy miles an hour.

The full story of "Cheltenham Flyer" is told in a book which takes its title from this famous train.*

* See "Cheltenham Flyer" (p. 109).

The Cornish Riviera

CORNISH RIVIERA LIMITED

DESPITE the reputation gained in the last century by such trains as "The Flying Dutchman" (11.45 a.m. Paddington to Exeter), it is doubtful if ever any train has attained more world fame than the "Cornish Riviera Limited" which leaves Paddington Station every weekday at 10.30 a.m. Its regular "non-stop" run from Paddington to Plymouth, a world's record when it was instituted in 1904, was maintained without challenge for nearly a quarter of a century with the exception of some adjustments due to national requirements during the war periods.

This train was also unique for many years in that, during the winter service, it carried three slip portions detached at speed at Westbury (for Weymouth), Taunton (for Minehead and Ilfracombe) and Exeter. Additional stops are now made at Exeter (Winter only) and Par (for Newquay).

In the Summer, when the Exeter stop is omitted, the

on Dainton Incline.

time for the $225\frac{3}{4}$ miles is four hours exactly, which gives an average speed of $56\frac{1}{2}$ miles per hour.

The "Cornish Riviera Limited" has always exemplified what is latest and best in rolling stock and appointments. Its general make-up is 14 coaches, each 60 ft. long over the corners, and 9 ft. 7 ins. wide—the widest railway coaching stock in this country. The train has seating for 464 passengers and 88 diners at a sitting.

This "aristocrat of trains," as it has been frequently called, is drawn by a "King" class engine as far as Plymouth, and the way in which engine power has had to be increased to conform with modern traffic requirements is illustrated by the fact that when the train was introduced in 1904 it was hauled by a "City" class (4-4-0) locomotive with a tractive effort of 17,790 lbs. and a weight (engine and tender) of 92 tons 1 cwt., whereas the comparable figures for the "King" class (4–6–0) engines are 40,300 lbs. and 135 tons 14 cwts. It must be remembered, however, that whereas in 1904 the weight of the seven-coach train (behind the tender) was 182 tons irrespective of passengers and luggage, that of the modern "Riviera" is 468 tons.

G.W.R. Express near Reading—Locomotive "King Henry IV."

G.W.R. Express passing Teignmouth, South Devon.

RAILWAY BOOKS for BOYS of ALL AGES

RAILWAYS, locomotives, and trains still hold high place in the affections of a host of young admirers notwithstanding all the fascination of aeronautics, wireless telephony, and other marvels of the wonderful age in which they live.

Young people realise that railway travelling is the easiest, safest, most expeditious, and luxurious system of transport. They recognise the influence of charm and wonder of the transit ways of steel and speed. And they demand information.

Conspicuous among a variety of attractive publications issued by the Great Western Railway, is a series of books which, whilst of considerable interest and educational value to the general reader, has been designed to appeal more particularly to those who are inquisitive as to the "hows" and "whys" of modern railway practice.

There are three books now available in this "Boys of All Ages" series which have been prepared primarily for the enlightenment and edification of young people, and these are described in the following pages.

They constitute wonderful value at one shilling each, having, on the average, 240 pages of descriptive matter containing 150 illustrations principally in half-tone, and a frontispiece in full colour or photogravure.

Some idea of the popularity of this series may be formed from the statement that about 185,000 books have already been distributed.

The publications of the Great Western Railway are obtainable from G.W.R. Stations or Offices, Railway Bookstalls, Agencies, Booksellers, or direct from the Superintendent of the Line, G.W.R., Paddington Station, London, W.2, from whom may also be obtained full particulars of train services and reduced fare travel facilities.

Photogravure frontispiece

260 pages
168 illustrations

Price 1/-

TRACK TOPICS

A Book of Railway Engineering for Boys of all Ages

By W. G. Chapman.

This book is concerned with the railway track and its structures—tunnels, cuttings, embankments, viaducts, bridges, etc.

The text consists of twenty descriptive "talks" which explain the problems confronting Brunel, the famous first Engineer of the Great Western Railway, in making his railway to the West, modern methods of construction and maintenance of the steel highway, and how it is taken over and under roads and rivers, through hills and across valleys, etc.

Get on the track of Track Topics

Locos: Old and New.

THE FAMOUS G.W.R. JIGSAW PUZZLES

(all of three-ply wood)

At 5/- each—about 400 pieces.

Drake Goes West—Plymouth, 1572.
Oxford—Brasenose College.
Stratford-on-Avon.

At 2/6 each—about 200 pieces.

G.W. Locos : In the Making.
Model Railway.
Romans at Caerleon.
Streamline Way.
Cornwall—Preparing for a Catch.
Historic Totnes.
Locos : Old and New.
Royal Route to the West.
Windsor Castle from the Air.

PRINTED IN GREAT BRITAIN BY CHELTENHAM PRESS LTD., CHELTENHAM AND LONDON

ENGINES

NAMES, NUMBERS, TYPES, CLASSES, ETC.

of

GREAT WESTERN RAILWAY LOCOMOTIVES

by

W. G. CHAPMAN

Published in 1946 by the

GREAT WESTERN RAILWAY

JAMES MILNE
General Manager

Paddington Station
London, W.2

PRICE TWO SHILLINGS & SIXPENCE

First Edition	*(present*	*series)*	*published*		1938
Second	,,	,,	,,	,,	1938
Third	,,	,,	,,	,,	1939
Fourth	,,	,,	,,	,,	1946

FOREWORD

THE first edition of this book (in its present form) was published in 1938. A reprint was demanded later in the same year, and yet another in 1939. Then came the second world war (1939–45) during the early part of which all stocks of the book were exhausted.

A steady demand for *G.W.R. Engines* persisted, but war conditions made any idea of reprinting quite impracticable. With the cessation of hostilities requests for information about G.W.R. locomotives were renewed, and it became evident that the interest evinced in the subject by large numbers of railway enthusiasts had in no way diminished.

In May, 1945, the *Great Western Railway Magazine* resumed its notes about new engines constructed, old engines condemned, etc., which had for many years been a regular feature of the journal, but which, for security reasons, had been discontinued in 1940.

It was soon apparent that the appetite for all sorts of facts and figures about G.W.R. locomotives could only be assuaged by the complete revision of *G.W.R. Engines*, and, just as soon as restrictions imposed by the scarcity of paper and labour permitted, this revised (fourth) edition was put in hand.

The opportunity has been taken to include a number of new features which it is felt will be welcomed by G.W.R. locomotive enthusiasts.

July, 1946. W. G. C.

Might and Majesty—a " King " Class Locomotive.

CONTENTS

Locomotive "King George VI," No. 6028, named after the reigning monarch, by His Majesty's gracious permission.

NAMING OF LOCOMOTIVES

THE naming of railway locomotives goes back beyond the opening of the first public railway in September, 1825, when George Stephenson's famous engine, *Locomotion*, astonished the world and introduced a new era of transport. Stephenson had himself built an engine named *Blucher* for work at the Killingworth Collieries eleven years before, and that early production was no isolated example of the naming of locomotives.

The first engines supplied to the Great Western Railway were built by outside contractors. All these bore distinctive names, the selection of which was apparently left to the makers. The first two engines delivered were *Vulcan* (Tayleur & Company of the Vulcan Foundry, Warrington) and *Premier* (Mather Dixon & Company, Liverpool). They were transported together by sea from Liverpool to London and thence by river and canal to West Drayton, where the first G.W.R. engine-house had been erected, and there delivered to the Company in November, 1837.

The selection of the name *Vulcan* is obvious, and *Premier* was apparently so called in the hope that it would be the first engine supplied to the Great Western Railway, although fate decreed that it should share that honour with another.

Although earlier locomotives acquired by the Great Western Railway, both passenger and goods,* were named but not numbered, it has been the practice for many years for all locomotives to be numbered and for passenger (tender type) engines to be named.

There was method in the allocation of names to G.W.R. locomotives from the start, for of the first twenty or so engines ordered by Brunel (the famous first engineer of the

* The first lots of engines built specifically for goods services were those of the "Leo" class (18 engines) 2–4–0 type, delivered 1841/2, and the "Hercules" class (4 engines) 0–6–0 type, delivered in 1842.

Railway), four built by Robert Stephenson & Company of Newcastle were *North Star, Morning Star, Evening Star,* and *Red Star;* engines from the Vulcan Foundry were *Vulcan, Aeolus, Bacchus, Apollo* and *Neptune;* those from Mather Dixon & Company were *Aerial, Ajax, Mercury,* and *Mars,* and other makers followed this practice of having associated names for their products.

As further engines were required by the Company they were ordered from various makers to designs prepared by Daniel Gooch, the Locomotive Superintendent, who had been appointed in August, 1837, when hardly twenty-one years of age, by Brunel. The first of these engines, *Firefly,* built by Jones, Turner & Evans of Newton, Lancashire, and delivered in March, 1840, supplied the class name for 62 engines with 7 ft. driving wheels supplied by a number of makers, all of whom adopted associated names for their engines, *e.g.*:—

FIREFLY CLASS

Builder	*Typical Names*
Jones, Turner & Evans, Newton	*Firefly, Wildfire, Spitfire.*
Sharp, Roberts & Co., Manchester	*Tiger, Leopard, Panther.*
Fenton, Murray & Jackson, Leeds	*Charon, Cyclops, Cerberus.*
Longridge & Co., Bedlington ..	*Jupiter, Saturn, Mars.*
Nasmyth, Gaskell & Co., Manchester.	*Achilles, Milo, Mentor.*
Stothert & Slaughter, Bristol ..	*Arrow, Dart.*
G. & J. Rennie, Blackfriars, London.	*Mazeppa, Arab.*

This system of naming was also followed with the twenty-one engines of the "Sun" class (6 ft. driving wheels) from various makers, and it is, perhaps, interesting to notice that those built by Sharp, Roberts & Company (who had selected names such as *Tiger, Leopard,* etc., for their "Firefly" Class engines) were called *Gazelle, Antelope, Zebra,* etc., whilst those supplied by Stothert & Slaughter, Bristol, had such names as *Javelin, Lance* and *Stiletto*. The engines which gave the name to the class were built by R. & W. Hawthorn of Newcastle, and were named *Sun, Sunbeam, Eclipse,* and so forth.

Probably enough has been said to indicate the method of allotting names to the earliest engines on the Great Western Railway, but perhaps a better appreciation of the way in which the names were applied and served the purpose of record is furnished by this reproduction of the first two pages of an old volume at Swindon entitled " List of Broad Gauge Passenger and Goods Locomotives on the Great Western Railway in 1868." (See pages 10 and 11.)

This shows that there were by that time eight classes of passenger engines and six classes of goods engines in service. Each class, it will be observed, following the practice referred to, was given a " master " name, the individual engines bearing other names. This system of only *naming* the engines continued, with comparatively few exceptions, throughout the broad gauge era, but with the change of gauge most of the named engines disappeared.

In the case of narrow (now standard) gauge, it was not the practice in the earlier years to name the engines, but in 1854 the Great Western Railway took over a batch of engines from the Shrewsbury and Chester Railway, bearing numbers 1 to 35. As these were the first standard gauge engines to come on to the line, they were given similar Great Western numbers and this apparently started the numbering of narrow gauge engines. Later, a further batch of engines, obtained from the Shrewsbury and Birmingham Railway, carried the numbers up to 56. A few of the engines bore name-plates at the time of their acquisition by the Great Western Railway, but these were gradually removed, probably to prevent confusion with broad gauge locomotives which did not entirely disappear until 1892.

The first Great Western narrow gauge engine built at Swindon Works came out in 1855. It was numbered 56, and from that time the numbering of G.W.R. engines became standard practice.

Description of the Classes of the Great Western Railway Locomotive Engines.

PASSENGER Engines in thick type, thus—"**Abbot.**"

"WOLF" .... Class, Tank Engines, Driving Wheels 6ft. single.

NOTE.—*The following of this Class have 7ft. Driving Wheels, viz.: "Bright Star," "North Star," "Polar Star," "Red Star," "Rising Star," "Shooting Star," and "Orion."*

BOGIE .... Class, Tank Engines, Driving and Trailing Wheels 5ft. 9in. Coupled

NOTE.—*"Brigand" and "Corsair" have 6ft. Wheels Coupled.*

METROPOLITAN .... Class, Tank Engines, Driving and Trailing Wheels 6ft. Coupled.

"PRIAM"............ Class, require Tenders, Driving Wheels 7ft. single.

NOTE.—*"Witch" has 7ft. 6in. Wheels.* Note the Vulcan Ostrich Phlegethon [illegible] Achilles & [illegible] have been renewed as 6ft Coupled Hawthorn Class

"ALMA" .... Class, require Tenders, Driving Wheels 8ft. Single.

"ABBOT"............ Class, require Tenders, Driving and Trailing Wheels 7ft. Coupled.

"VICTORIA"...... Class, require Tenders, Driving and Trailing Wheels 6ft. 6in. Coupled.

"HAWTHORN" Class, require Tenders, Driving and Trailing Wheels, 6ft. Coupled.

Cyls 16 × 24 Boiler [illegible] × 10 [illegible] Long

GOODS Engines in thin type, thus—"Ajax."

"LEO" .... Class, Tank Engines, Driving and Trailing Wheels 5ft. Coupled.

BANKING .... Class, Tank Engines, all wheels 5ft. Coupled.

"FURY" .... Class, require Tenders, all wheels 5ft. Coupled, (16in. Cylinders).

"CÆSAR" .... Class, require Tenders, all wheels 5ft. Coupled, (17in. Cylinders).

NOTE.—The Total Mileage of Engines which have been renewed, is reckoned from the date of renewal.

1

Name of Engine.	Maker.	Class.	Date of Starting.	Date when Renewed.	Miles run: During Half-year ending June 30 1866	Miles run: Total.
Abbot	Stephenson	Abbot	June '55		10.836	357.968
Abdul Medjid	G.W.R. 5th lot	Victoria	Oct. '56		17.279	324.377
Acheron	Fenton	Priam	Jan. '42	July 66	12.838	12.838
Achilles	Nasmyth	[illegible]	June '41			489.670
Actæon	„	„	Dec. „	Aug '56		220.038
Æolus	Tayleur	Wolf	Nov. '37			214.916
Ajax	G.W.R. 1st lot	Fury	May '46		11.878	315.544
Alexander	„ 5th „	Victoria	Nov. '56		19.329	326.387
Alligator	„ 2nd „	Cæsar	July '48		6.414	335.075
Alma	Rothwell	Alma	Nov. '54		9.700	310.142
Amazon	G.W.R. 4th lot	„	Mch. '51		14.717	413.552
Amphion	„ 6th „	Cæsar	April '56		6.125	230.227
Antelope	Sharp	Wolf	Aug. '41		10.760	430.334
Antiquary	Stephenson	Abbot	July '55		19.173	341.949
Apollo	Tayleur	Wolf	Jan. '38			193.252
Aquarius	Rothwell	Leo	June '42		7.201	313.359
Arab	Rennie	Priam	April '41		7.724	412.047
Argo	G.W.R. 1st lot	Fury	July '46			250.457
Argus	Fenton	Priam	Aug. '42	May '60	18.479	75.237
Ariadne	G.W.R. 5th lot	Cæsar	Nov. '52		3.887	246.850
Aries	Rothwell	Leo	June '41		8.884	340.392
Arrow	Stothart	Priam	July „			472.570
Assagais	„	Wolf	Sept. „	Jan. '64	14.426	79.575
Atlas	Sharp	„	June '38	July '60	5.073	62.048
Aurora	Hawthorn	„	Dec. '40			434.077
Avalanche	Stothart	*Banking*	Feb. '46			375.556
Avon	G.W.R. 6th lot	Cæsar	June '57		4.472	203.870
Azalia	„ 1st „	*Metropolitan*	April '64		10.714	59.305
Avonside late Slaughter	Avonside & Hawthorn		Dec 65		16.747	16.788
Bacchus	G.W.R. 3rd lot	Fury	May '49		3.959	177.788
Balaklava	Rothwell	Alma	Dec. '34		15.184	322.809
Banshee	G.W.R. 6th lot	Cæsar	Sept. „		12.651	271.824
Bee	Vulcan Foundry	*Metropolitan*	July '62		6.504	59.015
Behemoth	G.W.R. 2nd lot	Cæsar	Mch. '48		4.225	307.802
Bellerophon	„ 1st „	Fury	July '46		6.364	305.966
Bellona	Nasmyth	Priam	Nov. '41			505.057
Bergion	G.W.R. 1st lot	Fury	Jan. '47		10.186	271.726
Bey	Kitson	*Metropolitan*	July '62		2.633	55.371
Bithon	G.W.R. 6th lot	*Banking*	Oct. '54		9.032	247.698
Boyne	„ 7th „	Cæsar	Aug. '57		16.649	246.031
Briarens	„ 1st „	Fury	Feb. '47		8.732	310.652
					316.028	

Among the earlier narrow gauge engines named, in addition to being numbered, were some of the famous 7 ft. "singles," including the following :—

No.	*Date Constructed*	*Name*
378	September 1866	Sir Daniel
380	,, ,,	North Star
381	October ,,	Morning Star
471	June 1869	Sir Watkin
55	September 1873	Queen
999	March 1875	Sir Alexander
1118	,, ,,	Prince Christian
1122	April ,,	Beaconsfield
1123	May ,,	Salisbury

With these comparatively few exceptions, narrow gauge engines were not named until 1892—the year of the abolition of the broad gauge—when the first of the converted 7 ft. 8 in. "singles" came into service. Originally "convertibles," built in 1891, these engines bore numbers only, but after conversion names were allotted. Simultaneously, four four-coupled "convertibles" became standard gauge engines, and were named *Gooch, Armstrong, Brunel* and *Charles Saunders,* after the first Locomotive Superintendent, his successor (1864–1877), the first Engineer, and the first Secretary and General Superintendent of the Great Western Railway, respectively. It was soon after this that the practice of naming *and* numbering passenger train engines became general, and in many cases the names of old broad gauge engines were again adopted.

Probably the general motive for naming locomotives sprang from a natural desire of the makers to give their products an individuality, as in the case of ships from very early times, and as is the practice to-day in the case of commercial aircraft. It is possible, too, that even in the early days of railways, the publicity value of the naming of locomotives was in some measure appreciated. It cannot be denied that the practice increases public interest in the performances of particular engines, and it has to be remembered that stage coaches which were using the roads of the country in large numbers at the time of the advent of the railways, carried distinctive names by which their services were advertised to the public.

The *selection* of names is, perhaps, a little more difficult to explain in its entirety, but whilst early engine names were (as has been seen) largely mythical, historical, or physical, there was also a tendency to commemorate prominent individuals not only associated with the railways themselves, but with events of the times. There is an early example of this in Stephenson's *Blucher* (built in 1814), when that name was much in the news, and G.W.R. examples are provided in the engines *Beaconsfield* and *Salisbury*, constructed in 1875. An interesting example of the naming of a class of engines is found in the " Krugers " which came into being about the time of the Boer War, and yet another in the " Barnums " which appeared when that great American showman was " the talk of the town."

Names selected from associations with the Railway are furnished by the class of engines known as the " Devons " (referred to later), and by another class which became known as the " Aberdares," having been built primarily for the South Wales coal traffic. The " Badmintons " derived their name from the Railway's association with the celebrated Hunt, while the " Bulldog " class obtained its name from a general impression of power ; the engines of this class being somewhat similar to the " Devons," with larger boilers.

At the beginning of the present century, with the vast increase in the Company's locomotive stock, it was necessary to look further afield for names for locomotives and the " Flower," " City," " Saint," the original " County " and other classes came into being. In the " Flower " class, there was a large number of engines carrying names such as *Begonia, Auricula, Anemone,* but also some engines of similar construction with a variety of miscellaneous names, and those of *Baden-Powell, Kitchener,* and *Ladysmith* provide clues to the time of their appearing.

The famous " *Cities*," introduced in 1903, may be said to have inaugurated a system of naming successive batches and types of engines, with the view, first, of securing ready identification and, later, of helping in a scheme of standardisation. The names appropriated to the " *Cities* " were chosen from cathedral cities through which the Company's line runs; moreover, the common word "City" indicated a class.

In the "Saint" class again, many individual engines bore the name "Saint," as *Saint Agatha, Saint Andrew,* and it is, perhaps, a little surprising to find in such pious company *Lady Godiva, Lady Disdain,* and *Lady Macbeth.* Besides a number of "*Ladies,*" the "Saint" class included a series of "*Courts*" and also some engines which took their names from the Waverley novels.

With the original "County" class, however, all engines in the class were named after counties in the British Isles. The first was *County of Middlesex,* the idea being to start at Paddington, which is in Middlesex, and proceed down the line through Berkshire, Wiltshire, etc., naming an engine after each county. Later on, the type having to be increased to a greater number than there were counties served by the Great Western Railway, other counties were selected.

In the "Star" class we find, as well as a number of engines bearing the names of stars, such as *Morning Star, North Star, Polar Star,* a distinguished company of "*Queens,*" "*Princes,*" "*Princesses,*" and others, for although the stars of the heavens "cannot be numbered," those suitable for engine names were apparently limited.

Coming to more modern times, we find the practice of adopting names of two or three words (as in the "Counties" and "Cities"), embodying a "master" name for a class of engines and an individual name for each locomotive, has been more or less strictly applied: the idea being that either the first or last word should be common to a class and the remainder of the name lend itself to alphabetical reference.

In the "Castle" class, the prototype of which was *Caerphilly Castle,* No. 4073, the vast majority of the engine names embody the word "Castle." There are, however, in this class twenty-one "*Earls*" (*e.g.* No. 5043 *Earl of Mount Edgcumbe*), nine "*Abbeys*" (*e.g.* No. 5083, *Bath Abbey*), twelve aircraft types (*e.g.* No. 5071, *Spitfire*) and ten others whose names do not embody the word "Castle." Certain engines of the "Star" class have been reconstructed as "Castles," including all but two* of the "*Abbey*" engines. These retain their name-plates and have been renumbered 5083 and 5092.

* No. 4061 *Glastonbury Abbey* and No. 4062 *Malmesbury Abbey.*

It was during recent war years that a number of " Castle " class engines were renamed after various types of Royal Air Force machines.

In the " King " class, all the engines have the names of Kings of England, the first engine which was produced in 1927 being named *King George V*, after the then reigning Sovereign. The other engines in the class go back to *King Richard I* (No. 6027). The name of our present King George VI is carried on engine No. 6028, and the name-plate *King Edward VIII* is on engine No. 6029.

In the " Hall," " Grange," and " Manor " classes, all two-cylinder 4–6–0 engines, the names are taken from country houses in Great Western Railway territory, with the single exception that " Hall " class engine No. 4900 is named " *Saint Martin*."

The advantages to the Railway of thus naming locomotives on systematic lines will be fairly apparent, particularly as the more precise adoption of this form of naming has been contemporary with the standardisation of engines and engine parts. It will not be difficult to appreciate that standardisation lends itself to nomenclature, and that the system has many advantages when, for example, it is known that a particular class of engine carries a certain standard type of cab, boiler, etc., and works with a certain standard tender.

There are, as might be expected, a few exceptions in the class system of naming, and these have in all cases been made for quite special reasons. There is, for example, in the " Star " class an engine numbered " 100 A1," and named *Lloyds*. This change was made as a complimentary gesture, as was the re-naming of engine No. 4037 *South Wales Borderers*. There are other cases, though the number in the aggregate is quite small.

Naming, in addition to numbering, engines is, of course, not really necessary. There is, however, a good deal to be said in favour of it, one valuable feature being that it serves as an aid to memory. Numbers are difficult to carry in mind, and it may fairly be said, so far as Great Western engines are concerned, that the names of individuals and of types constitute a means of reference of a distinctly useful character in many directions.

A "King" and a "Castle" en route to London

NUMBERS, TYPES, AND CLASSES

As has already been seen, the numbering of G.W.R. locomotives started in 1854 with the acquisition of engines from the Shrewsbury and Chester Railway, and the practice was generally established in the following year.

Numbers are not issued strictly consecutively as engines are constructed, but blocks of numbers are allocated to particular classes of locomotives. An example of the method of allocation of numbers is furnished by the " Hall " class. The numbers commence at 4900, run to 4999 and then proceed from 5900 to 5999 and then from 6900 onwards. In the same way in the "5700" class of 0–6–0T type engines the serial numbers are 57**, 67**, 77**, 87** and so on.

As only passenger (tender type) engines are named, it is not surprising to learn that engines are generally known on the railway by their class numbers ; for example, the "Kings" are the " 6000 " class, " Castles " the " 4073 " class, " Halls " the " 4900 " class, while goods and shunting engines (not named) are known similarly by their class numbers, *e.g.*, " 4300," " 2800," and "4500T " classes.

The practice will be made clear from the information given later in which each class of locomotives is separately described.

As an aid to memory the following table may be found useful in identifying particular types of locomotives from their number plates, but it should be explained that this table refers only to standard G.W.R. locomotives, and there are in the G.W.R. service engines of various descriptions taken over from absorbed and constituent railways bearing numbers to which the table is not applicable. All G.W.R. modern standard type steam locomotives have four-figure numbers with the following exceptions, and these are, of course, not covered by the table :—

100 A1	..	*Lloyd's*	Castle Class
111	..	*Viscount Churchill*	,, ,,

STANDARD TYPE G.W.R. LOCOMOTIVES

Tender Types

0–6–0 engines all numbered 22**, 23**, 24**, 25**.
4–4–0 ,, ,, ,, 32**, 33**, 34**

		Second figure in Engine Number
4 cylinder	4–6–0	0
2 ,,	4–6–0	0, 8 or 9
2 ,,	2–6–0	3 or 6
2 ,,	2–8–0	7 or 8

Tank Types (all 2 cylinder)

0–4–2T	.. .. ..	8
0–6–0T	.. .. ..	3, 4, 6, 7 or 9
2–6–2T	.. .. ..	1 or 5
(except 11 engines numbered 4400–4410)		
2–8–0T } 2–8–2T }	.. .. ..	2
(7200 class only).		
0–6–2T	.. .. ..	6

Locomotives are also classified by their wheel arrangements—4–6–0, 0–6–0T, etc. Under this system, the first digit indicates the number of wheels (pony truck or bogie) in front of the coupled wheels ; the second, or middle digit, represents the number of coupled driving wheels ; and the third digit

CLASSIFICATION OF ENGINES.

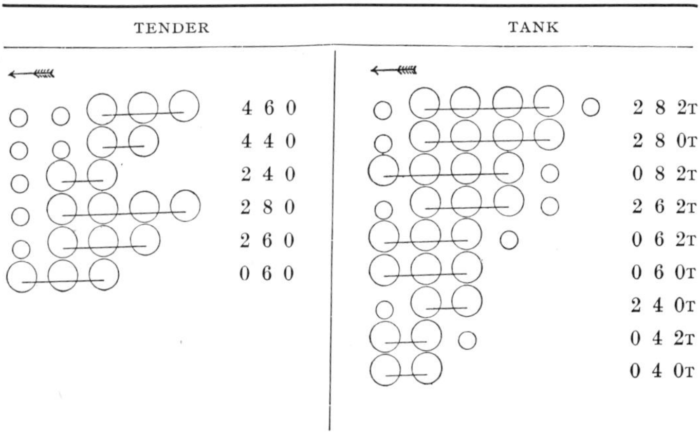

TENDER	TANK
4 6 0	2 8 2T
4 4 0	2 8 0T
2 4 0	0 8 2T
2 8 0	2 6 2T
2 6 0	0 6 2T
0 6 0	0 6 0T
	2 4 0T
	0 4 2T
	0 4 0T

signifies the number of wheels (trailing) behind the coupled wheels. Thus, in the 10-wheeled " King " class engine, there is the *four*-wheel bogie, *six*-coupled driving wheels and *no* trailing wheels—4–6–0. Tank engines are distinguished from tender types by the addition of the letter " T," thus 2–8–2T.

The following names, American in origin, are also used to designate classes of locomotives by wheel arrangement, but are not so much in favour in this country as they were some years ago :—

Atlantic	4–4–2
Pacific	4–6–2
Baltic	4–6–4
Mogul	2–6–0
Mikado	2–8–2
Prairie	2–6–2
Decapod	2–10–0
Consolidation	2–8–0

EXAMPLES OF LOCOMOTIVE CLASSIFICATION BY WHEEL ARRANGEMENT

Type	Size of Wheels	Classes = Passenger Engines Numbers = Goods, Mixed & Branch	Class
4–6–0	6′ 6″	4 cylinder " King " ..	" King "
	6′ 8½″	" Star "	" Star "
		" Knight "	
		" Monarch "	
		" Queen "	
		" Prince "	
		" Princess "	
		" Abbey "	
		" Castle " ..	" Castle "
4–6–0	6′ 8½″	2 cylinder " Lady "	" Saint "
		" Saint "	
		" Court "	
		Various	
	6′ 3″	" County " ..	" County "
4–6–0	6′ 0″	" Hall " ..	" Hall "
4–6–0	5′ 8″	" Grange " ..	" Grange "
4–6–0	5′ 8″	" Manor " ..	" Manor "
4–4–0	5′ 8″	" Director "	" Bulldog "
		" Place "	
		" Bird "	
		Various	
		" Duke " or " Devon "	" Duke "
		" 3200 " ..	" 3200 "

EXAMPLES OF LOCOMOTIVE CLASSIFICATION BY WHEEL ARRANGEMENT *(continued)*

Type	Size of Wheels	Classes = Passenger Engines Numbers = Goods, Mixed & Branch	Class
2-6-0	4′ 7½″	2600-2680	
	5′ 8″	4300-4399	
		5300-5399	
		6300-6399	
		7300-7321	
		9300-9319	
0-6-0	5′ 2″	2200-2299	
0-6-2T	4′ 7½″	5600-5699	
		6600-6699	
0-6-0T	5′ 2″	5400-5424	
	4′ 7½″	5700-5799	
		6700-6749	
		7400-7429	
		7700-7799	
		8700-8799	
		9711-9799	
		3700-3784	
		9700-9710 Condenser	
		6400-6439	
	3′ 8″	1366-1371	
2-6-2T	5′ 8″	3150-3190	
		4100-4129	
		5100-5199	
		6100-6169	
	4′ 7½″	4500-4599	
		5500-5574	
	4′ 1½″	4400-4410	
2-8-0	5′ 8″	4700-4708	
	4′ 7½″	2800-2883	
2-8-0T	4′ 7½″	4200-4299	
		5200-5254	

G.W.R. locomotives are also classified for "routing" purposes. The permanent way is divided into sections according to the carrying capacity of bridges, etc., and these sections are designated red, blue, yellow, and uncoloured routes. The routes over which a particular engine may work are indicated by means of coloured discs (or the absence of such discs) on the sides of the engine cabs.

Most of the main line consists of red routes over which all engines may run, the heavy engines carrying a red disc being restricted to these routes. Blue routes may be traversed by engines carrying blue or yellow discs, or no disc at all, but not by engines carrying a red disc. Yellow routes are restricted to engines carrying a yellow disc or no disc at all, and uncoloured routes are restricted to engines carrying no disc indications.

"King" class engines, being the heaviest of all, carry two red discs and are restricted to specified portions of the red routes.

There is a further classification of locomotives for "power" purposes, and this is indicated by a letter superimposed upon the coloured disc on the cab sides of the engines.

Most of the engines are thus classified, although those of low tractive power are unlettered. The lettered engines are placed in groups A, B, C, D, and E, according to their hauling capacity—the E group being the most powerful.

A line of "Kings"—Swindon Running Shed

HISTORICAL SURVEY

IT is not proposed to trace the development of Great Western Railway locomotives through last century, for that has already been done in some detail elsewhere.

Something has already been said about the earliest engines, and it may be of interest to give a brief chronological survey of the types introduced from 1895 onwards.

1895 Locomotives of the 4–4–0 type, with 5 ft. 8 in. driving wheels, for working over heavy inclines in the West of England, were introduced. They were known as the "Devons," but as the first in the series was named *Duke of Cornwall*, the class came to be familiarly referred to as the "Dukes." Incidentally, it may be remarked that *Duke of Cornwall*, No. 3252, only ceased work in September, 1937.

1896 The first 4–6–0 type engine (No. 36) was built for heavy goods work. This engine had 4 ft. 7½ in. driving wheels and 20 in. by 24 in. cylinders.

1897 The "Badmintons," 4–4–0 type with inside cylinders and motion like the "Devons," appeared and were forerunners of the "Cities."

1898 The fleet of "Devons" was augmented, and one of the last twenty was fitted with the standard No. 2 boiler and called *Bulldog* (No. 3311)—thus the "Bulldog" class evolved. Commencing with No. 3341, the outside frames, instead of being cut down between the driving and trailing wheels, were continued straight through. In this form, and with the exception that the boiler barrel is coned, a large number of these engines has since been built.

1900 A further development of the "Badmintons," with straight frames—the "Atbaras"—were turned out, being somewhat similar to the "Bulldogs" and having standard No. 2 boilers.

In this year also, the first of the "Aberdares," 2–6–0 class, was built. They had coupled wheels 4 ft. 7½ in. diameter. More were built the following year, and since that time the number has been increased, but they have now been fitted with No. 4 boilers.

Types of Locomotives Introduced 1895 to 1901

(Mr. William Dean, G.W.R. Locomotive Superintendent)

P—Passenger. **G**—Goods. **M**—Mixed Traffic. **B**—Branch. (See last column.)

Date.	CLASS.		Type.	CYLINDERS.		Boiler Pressure lbs. per sq. in.	Diameter of Coupled Wheels.	
	Name.	No.		No.	Diam. and Stroke.			
					in. in.		ft. in.	
1895	Duke	3252	4-4-0	2 ins.	18 × 26	165	5 8	**P**
1896		36	4-6-0	2 ins.	20 × 24	165	4 7½	**G**
1897	Badminton	4100	4-4-0	2 ins.	18 × 26	180	6 8½	**P**
1898	Bulldog	3312	4-4-0	2 ins.	18 × 26	165	5 8	**P**
1899	Kruger	2601	4-6-0	2 ins.	19 × 28	180	4 7½	**G**
1900		2600	2-6-0	2 ins.	18 × 26	200	4 7½	**G**
1900	Atbara	4120	4-4-0	2 ins.	18 × 26	180	6 8½	**P**
1901		2602	2-6-0	2 ins.	19 × 28	165	4 7½	**G**
1901		3600	2-4-2T	2 ins.	17 × 24	195	5 2	**M**

1902 This year saw the introduction of the first of the 4–6–0 2-cylinder type (No. 100). After running some months the engine was named *William Dean*, being re-numbered 2900 some years afterwards. The boiler had a straight barrel, but later, and on succeeding engines, the barrel was coned.

1903 A new class of 4–4–0 locomotives, the "Cities," which was destined to make history, was introduced. It was an engine of this class which set up a speed record for a railway train which stood unchallenged for over thirty years.

At this time also, the first of the 2–8–0 type (No. 97, now 2800) goods engines was built, and the first of the 2–6–2T class (No. 99—now 5100) put into traffic. The De Glehn compound engine, *La France*, was purchased during this year. Subsequently, two more powerful engines of this type were added, and appropriate names—*President* and *Alliance*—were chosen for them, their acquisition synchronising with the launching of " L'Entente Cordiale." Although not of G.W.R. standard, they were subsequently brought within the " interavailability-of-boiler " scheme, and provided with G.W.R. boilers.

1904 In this year, the first of the " Counties " (No. 3473, later 3800), 4–4–0 type, was built. Possessing features peculiar to the class, the scheme of naming followed in the " Cities " was repeated, the names being selected from the counties served by the Railway.* .

1905 Six more 2-cylinder 4–6–0's, similar to the sample built in 1902, were constructed, and five were eventually given the names of certain members of the Great Western directorate. An entirely new type (4–4–2 " Atlantics "—outside cylinders) appeared, to which names borne by the " Waverley " class of broad-gauge engines—*Lalla Rookh, Redgauntlet, Robin Hood, Rob Roy, The Abbot,* etc., were given. They were later converted to 4–6–0's.

This did not complete the activities of the year, as the " County Tanks," (4–4–2T) type, were introduced.

1906 The first of the 4-cylinder engines (No. 40, now 4000) came into service. Having regard to the engine's exceptional design, it was decided to perpetuate upon it the name *North Star* borne by famous predecessors.

Subsequently, engines of this type were also given names in succession to the broad-gauge " Star " series. The first engine only, No. 40, was built experimentally with a 4–4–2 wheel arrangement, but was later altered to 4–6–0 type, many of which are now in service.

Next came a batch of 2-cylinder 4–6–0's, which were given names commencing with " *Lady* " after certain famous characters in fiction.

1907 Further engines of the 2-cylinder 4–6–0 type were built, possessing one or two distinctive features, to which the names of *Saints* were allotted.

* *See* Notes for Year 1945 (page 32).

1908 In June, the first engine of the "Pacific" type (4–6–2) built in this country, *viz.*, No. 111, *The Great Bear*, was completed. This engine was a development of the "Stars"; the cylinders, motion, expansion gear, axles and bogies being similar.

Another batch of engines similar to the "Stars" followed and was given names indicative of Orders of Knighthood.

A new type of 4–4–0 inside cylinder engines with 6 ft. 8½ in. driving wheels was also introduced this year—given the names of flowers, thus individualising the class.

1909 This type was followed by a "lot" of 5 ft. 8 in. 4–4–0 inside cylinder engines, with names of birds, the "Bird" class signalising a temporary return to the 5 ft. 8 in. 4–4–0 inside cylinder type.

More 4–6–0 4-cylinder engines, similar to the "Stars" and "Knights," appeared, and it was decided to name these after certain of the Kings of England. The first was *King Edward*.

1910 The 2–8–0T type ("4200" class for heavy goods traffic) was introduced, followed by a further batch of 4–6–0–4-cylinder engines, and named after the Queens of England, *Queen Mary* being the first.

1911 This year witnessed the production of the first 2–6–0 ("4300" class) outside cylinder engines of which an important feature was the interchangeability of all parts with the 2–6–2T type.

An additional number of "*Saints*" and "*Ladies*" were constructed, and named after historical residences; these became known as the "*Courts*."

1913 More "Star" class engines were built, commencing with *Prince of Wales*; they were named after the Princes of the reigning Royal Family.

1914–1918 There was no exceptional development during this War period, standard types of engines being produced.

1919 In May, a new 2–8–0 ("4700" class) engine was introduced for dealing with heavy express mixed traffic. This engine had a standard No. 7 boiler, and the same wheel arrangement as the standard " 2800 " class, but instead of the 4 ft. 7½ in. diameter driving wheels of that class, it had driving wheels of 5 ft. 8 in. diameter.

Types of Locomotives Introduced 1902 to 1919

(Mr. G. J. Churchward, G.W.R. Chief Mechanical Engineer)

P—Passenger. **G**—Goods. **M**—Mixed Traffic. **B**—Branch. (See last column).

Date	CLASS.		Type.	CYLINDERS.		Boiler Pressure lbs. per sq. in.	Diameter of Coupled Wheels	
	Name.	No.		No.	Diam. and Stroke.			
					in. in.		ft. in.	
1902	William Dean	2900	4-6-0	2 outs.	18 × 30	200	6 8½	**P**
1903		2611	2-6-0	2 ins.	18 × 26	200	4 7½	**G**
1903		3100	2-6-2T	2 outs.	18 × 30	195	5 8	**M**
1903	City of Bath	3710	4-4-0	2 ins.	18 × 26	195	6 8½	**P**
1903	Ernest Cunard	2998	4-6-0	2 outs.	18 × 30	200	6 8½	**P**
1903	La France (purchased)	102	4-4-2	4 {2 hp. 2 lp. comp'nd	$13\frac{3}{8} \times 25\frac{3}{16}$ $22\frac{1}{16} \times 25\frac{3}{16}$	227	6 8½	**P**
1903	Albion	2971	4-6-0	2 outs.	18 × 30	200	6 8½	**P**
1904	Albion converted to	2971	4-4-2	2 outs.	18 × 30	200	6 8½	**P**
1904	County of Middlesex	3800	4-4-0	2 outs.	18 × 30	200	6 8½	**P**
1904		4400	2-6-2T	2 outs.	16½ × 24	165	4 1½	**G**
1905	Alliance and President (purchased)	103 & 104	4-4-2	4 {2 hp. 2 lp. comp'nd	$14\frac{3}{16} \times 25\frac{3}{16}$ $23\frac{5}{8} \times 25\frac{3}{16}$	227	6 8½	**P**
1905		2801	2-8-0	2 outs.	18 × 30	200	4 7½	**G**
1905	County Tank	2221	4-4-2T	2 outs.	18 × 30	195	6 8½	**P**
1905	Ivanhoe	2981	4-4-2	2 outs.	18 × 30	225	6 8½	**P**
1906	Krugers converted to	2601	2-6-0	2 ins.	18 × 26	200	4 7½	**G**
1906	Lady Superior	2901	4-6-0	2 outs.	18 × 30	225	6 8½	**P**
1906	North Star	4000	4-4-2	4	14 × 26	225	6 8½	**P**
1906		4500	2-6-2T	2 outs.	17 × 24	180	4 7½	**G**
1907		2821	2-8-0	2 outs.	18⅜ × 30	225	4 7½	**G**
1907	Albion reconverted to	2971	4-6-0	2 outs.	18 × 30	225	6 8½	**P**
1907	Dog Star	4001	4-6-0	4	14¼ × 26	225	6 8½	**P**
1907	25 –'s converted to	3901	2-6-2T	2 ins.	17½ × 24	180	5 2	**M**
1907	Prairie Tank	3151	2-6-2T	2 outs.	18½ × 30	200	5 8	**M**
1908	The Great Bear	111	4-6-2	4	15 × 26	225	6 8½	**P**
1908	Auricula (Flowers)	4101	4-4-0	2 ins.	18 × 26	195	6 8½	**P**
1909	Blackbird	3441	4-4-0	2 ins.	18 × 26	195	5 8	**P**
1909	North Star converted	4000	4-6-0	4	14¼ × 26	225	6 8½	**P**
1910	Consolidation Tank	4201	2-8-0T	2 outs.	18½ × 30	200	4 7½	**G**
1911		2831	2-8-0	2 outs.	18½ × 30	225	4 7½	**G**
1911		4301	2-6-0	2 outs.	18½ × 30	200	5 8	**M**
1912	Ivanhoe converted to	2981	4-6-0	2 outs.	18 × 30	225	6 8½	**P**
1913		4530	2-6-2T	2 outs.	17 × 24	200	4 7½	**G**
1913		4600	4-4-2T	2 outs.	17 × 24	200	5 8	**B**
1913	Prince of Wales	4041	4-6-0	4	15 × 26	225	6 8½	**P**
1919		4700	2-8-0	2 outs.	19 × 30	225	5 8	**M**

1923 In August, *Caerphilly Castle* (No. 4073), the first of the " Castle " class of engines, was put into service. This class of engine having proved very successful, it was decided to build more of them. In addition, five of the older 4-cylinder engines in the " Star " class were subsequently converted to " Castle " class.

1924 *The Great Bear* (No. 111), the only Great Western engine of the " Pacific " type, was reconstructed as a " Castle " and re-named *Viscount Churchill*. In December a new type of engine—0–6–2T (" 5600 ")—specially designed to deal with the heavy traffic of the South Wales coalfields over the long gradients and sharp curves met with in that area, was put into service.

In December also, one of the 4–6–0 class, No. 2925, *Saint Martin*, was rebuilt with coupled wheels of 6 ft. 0 in. diameter instead of 6 ft. 8½ in., and fitted with a cab similar to the " Castle " class.

1927 In July the famous engine *King George V* (No. 6000), the first of the " King " class of locomotives, was put into service. These engines, the most powerful of the passenger type in Great Britain, were specially designed to cope with the heavy traffic and fast running on certain sections of the line, and embody several special features.

With the advent of the " Kings," the " Star " class engines bearing the names of " Kings " were renamed " Monarchs." The name-plates were also removed from other odd engines bearing the names of kings.

1928 Additional engines of the " King " class, and a new series of 4–6–0 engines, the " Halls " (" 4900 "), were constructed : the latter being named after well-known halls on the G.W.R. system.

1929 This year saw the advent of a new series of 2–6–2T engines (" 5100 " class) which were specially built for suburban passenger services, also a new class of 0–6–0T engines (" 5700 ") designed for shunting purposes.

1930 An additional class of 0–6–0 engines (" 2251 " class) was constructed for light mixed traffic.

1931–1933 Auto-engines of the 0-6-0T type ("5400" and "6400" classes) were built during 1931 and 1932, and others of the 0–4–2T type (" 4800 " class) during 1932 and 1933.

1933 A further batch of 0–4–2T type (" 5800 " class) not fitted with auto-gear were constructed.

Eleven condensing engines of the 0–6–0T type (" 9700 " class) were constructed to replace the older type working over the Metropolitan lines of the London Underground.

1932–
1934 Additional engines of the 4–6–0 type (" Hall " and " Castle " classes); 2–6–0 type (" 9300 " class); 2–6–2T type (" 6100 " class), and 0–6–0T type (" 8750 " class), were built.

1934 A batch of six small shunting engines of the 0–6–0T type (" 1366 " class) was put into service in March, 1934—replacing older engines of the same type. Later in the year, twenty engines of the 2–8–0T type (numbered 5275–5294), originally specially designed for dealing with the coal traffic from pit to port, but now no longer required for this service owing to the falling-off in the coal export trade, were converted into 2–8–2T type (a new departure for the G.W.R.) to make them suitable for main line freight train service, and numbered 7200 to 7219.

1935 A further batch of twenty engines (numbered 5255–5274) was converted to 2–8–2T type and numbered 7220–7239.

1936 In May the first of the " 3200 " class, 4–4–0 type, appeared. These engines replaced the older " Duke " and " Bulldog " classes for light axle-load routes.

In August the first of the "7400" class, 0-6-0T type, shunting engines appeared, with $16\frac{1}{2} \times 24$ in. cylinders and 4 ft. $7\frac{1}{2}$ in. wheels, followed in September by the first of the "Grange" class, 4-6-0 type, having $18\frac{1}{2}$ in. × 30 in. cylinders and 5 ft. 8 in. wheels. Designed for dealing with fast freight and passenger services, these engines replaced the "4300" class, 2–6–0 type.

1937 Various engines of standard types built.

1938 The first of the " Manor " (" 7800 " class), 4–6–0 type, appeared early in the year, having 18 in. × 30 in. cylinders and 5 ft. 8 in. driving wheels. These engines are for work similar to that performed by the "Grange" class, but are lighter in weight to allow of them being used over lines where heavy engines are not permitted.

Locomotive No. 2872 : One of ten engines of "2800" class converted to burn oil fuel instead of coal ; other engines similarly adapted were Nos. 2832, 2839, 2849, 2854, 2862, 2863, 2888, 3818 and 3865. These will eventually be renumbered under the 4800 group. 0–4–2T engines now in the 4800 class will be transferred to a 1400 group.

In March a new batch of 2–8–0 type engines (" 2884 " class) were produced having larger and improved cabs.

Some of the " 5101 " class 2–6–2–T type engines were fitted with 5 ft. 6 in. wheels and the boiler pressure increased to 225 lbs. per sq. in. These were re-numbered " 8100 " class. Later some of the " 3150 " class, 2–6–2T type engines, were fitted with 5 ft. 3 in. wheels and boiler pressure increased to 225 lbs. per sq. in. These were re-numbered " 3100 " class.

Types of Locomotives Introduced 1923 to 1941

(Mr. C. B. Collett, G.W.R. Chief Mechanical Engineer)

P—Passenger. **G**—Goods. **M**—Mixed Traffic. **B**—Branch. (See last column).

Date.	CLASS.		Type	CYLINDERS.		*Boiler Pressure lbs. per sq. in.*	*Diameter of Coupled Wheels.*	
	Name.	No.		No.	Diam. and Stroke.			
					in. in.		ft. in.	
1923		5205	2-8-0T	2 outs.	19 × 30	200	4 7½	**G**
1923	Caerphilly Castle	4073	4-6-0	4	16 × 26	225	6 8½	**P**
1924	The Great Bear converted to Castle and renamed Viscount Churchill	111	4-6-0	4	16 × 26	225	6 8½	**P**
1924		5600	0-6-2T	2 ins.	18 × 26	200	4 7½	**M**
1925	Saint Martin rebuilt	2925	4-6-0	2 outs.	18½ × 30	225	6 0	**P**
1927	King	6000	4-6-0	4	16¼ × 28	250	6 6	**P**
1928	Hall	4901	4-6-0	2 outs.	18½ × 30	225	6 0	**P**
1929		5700	0-6-0T	2 ins.	17½ × 24	200	4 7½	**G**
1929		5101	2-6-2T	2 outs.	18 × 30	200	5 8	**M**
1930		2251	0-6-0	2 ins.	17½ × 24	200	5 2	**M**
1931		6100	2-6-2T	2 outs.	18 × 30	225	5 8	**M**
1932		5400	0-6-0T	2 ins.	16½ × 24	165	5 2	**B**
1932		6400	0-6-0T	2 ins.	16½ × 24	165	4 7½	**B**
1932		4800	0-4-2T	2 ins.	16 × 24	165	5 2	**B**
1933		9700	0-6-0T	2 ins.	17½ × 24	200	4 7½	**G**
1934		1366	0-6-0T	2 ins.	16 × 20	165	3 8	**G**
1934		7200	2-8-2T	2 outs.	19 × 30	200	4 7½	**G**
1936	Grange	6800	4-6-0	2 outs.	18½ × 30	225	5 8	**M**
1936		3200	4-4-0	2 ins.	18 × 26	180	5 8	**P**
1936		7400	0-6-0T	2 ins.	16½ × 24	180	4 7½	**G**
1938	Manor	7800	4-6-0	2 outs.	18 × 30	225	5 8	**M**
1938		2884	2-8-0	2 outs.	18½ × 30	225	4 7½	**G**
1938		8100	2-6-2T	2 outs.	18 × 30	225	5 6	**M**
1938		3100	2-6-2T	2 outs.	18½ × 30	225	5 3	**M**

1939–
1944 There were no exceptional developments during the War period ; standard types of engines being produced, but in 1944 a batch of improved " Hall " class engines, 4–6–0 type, appeared, numbered 6959–6970. These were fitted with straight-through frames

and plate frame bogies. A new type of triple row superheater was also fitted in the boilers of the first seven engines.

1945 In August the first engine of the "1000" class, (4–6–0) type, appeared with $18\frac{1}{2}$ in. × 30 in. cylinders, 6 ft. 3 in. wheels, and a boiler pressure of 280 lbs. per sq. in. The first engine only was fitted with a double chimney and blast pipe. A new pattern straight-sided tender is in use with this class.

The original "County" class of locomotives introduced in 1904 having all been condemned (the last in 1933), "County" names are being applied to "1000" class engines (*see* p. 50).

Certain engines of the "2800" class were converted to burn oil fuel instead of coal.

Types of Locomotives Introduced 1941—1945.
(Mr. F. W. Hawksworth, G.W.R. Chief Mechanical Engineer)

Date.	CLASS.		Type.	CYLINDERS.		Boiler Pressure lbs. per sq. in.	Diameter of Coupled Wheels.	
	Name.	No.		No.	Diam. and Stroke.			
					in. in.		ft. in.	
1944	Hall	6959	4-6-0	2 outs.	$18\frac{1}{2} \times 30$	225	6 0	M
1945	County	1000	4-6-0	2 outs.	$18\frac{1}{2} \times 30$	280	6 3	M

The tables included above give the years in which new types of engines were first constructed under the régime of successive Locomotive Superintendents. These tables show the number of cylinders, diameters, boiler pressures, and diameters of coupled wheels, thus briefly summarising G.W.R. locomotive history over each period.

GREAT WESTERN RAILWAY
Locomotive Superintendents and Chief Mechanical Engineers

1837–1864	..	Sir Daniel Gooch
1864–1877	..	Joseph Armstrong
1877–1902	..	William Dean
1902–1921	..	George J. Churchward
1921–1941	..	Charles B. Collett
1941–	..	F. W. Hawksworth

During Mr. Churchward's régime the title of Locomotive Superintendent was changed to that of Chief Mechanical Engineer.

The "Torbay" Express at Dawlish—Locomotive "King Henry III"

Birmingham-Paddington Express. Locomotive No. 6026 "King John"

STANDARD LOCOMOTIVE CLASSES AND TYPES

THE following section embraces the existing standard types of G.W.R. engines, but it should be explained that as a result of amalgamations, there are many engines still in service, distributed over the system, wichh do not fall into any of these classes, and also some older types of G.W.R. engines.

The general design of a locomotive is governed by the duties it has to perform : express passenger engines, such as those of the " King " and " Castle " classes, are built with driving wheels of large diameter in order to obtain high speeds with a normal piston speed. This high speed is, however, only gained at the loss of tractive effort, as may be seen when comparing express passenger engines with the " 2800 " class goods engines. The driving wheels of the latter are smaller in diameter, but the tractive effort is comparatively high, these engines being used for heavy freight trains, where high speed is not of paramount importance.

When comparatively high speeds are required with a fair tractive effort a compromise is effected as in the " mixed-traffic " engines, the wheel diameter being designed to give just the required speed and no more in order to keep the tractive effort as high as possible.

In the following pages tender type engines are followed by tank types. The former are necessary, especially on long " non-stop " runs, as the water and fuel capacity of the tank type locomotive is not adequate for this class of work. The function of the tender is to carry the fuel, water, oil, tools, etc., required on the locomotive. The tender also embraces the water pick-up apparatus and is braked both by vacuum and hand. The tender being connected to the engine, the whole forms a flexible power unit.

G.W.R. Tender

In the following table particulars are given of the types of locomotive tenders in use on the Great Western Railway, together with their water and coal capacities, their weights full and empty, and the classes of engines to which each type of tender is fitted.

Water capacity Gallons	Coal capacity Tons	Coal capacity Cwts.	Weight full Tons	Weight full Cwts.	Weight empty Tons	Weight empty Cwts.	Classes of Locomotives to which fitted
2,500	4	10	34	5	16	2	2301
3,000	5	0	36	15	17	9	2251, 2301 2600, 3300
3,000	4	5	39	15	22	2	2251
3,500	7	0	40	0	18	5	2800, 2900 3200, 3300 4300, 6800 7800.
4,000	6	0	46	14	22	10	4000, 4073 4700, 4900 6000.
4,000	7	0	49	0	22	14	1000.
R.O.D. 4,000	7	0	47	14	23	0	2251, 2600, 3000 (R.O.D.)

Tank type locomotives are required where it is necessary to be able to drive in either direction without the use of a turn-table. Tank engines are also used for shunting operations, branch line work, and generally where the loads dealt with do not justify the use of the larger tender-type locomotive.

The use of the G.W.R. monogram on the tenders or tanks of locomotives (as on page **36**) and on the large majority of engines here depicted is being discontinued.

The new standard markings will be :—

For named engines

The letters " G. . . W " with the G.W.R. coat-of-arms between the two letters, as on page 50 (" County Class ").

For all other engines

The letters " G.W.R " as on page 30 (" 2800 " Class).

The old monogram will be deleted and the new marking imposed as locomotives pass through the shops for repairs or periodical overhaul.

3300 ("Bulldog") Class. Type 4—4—0

(Introduced 1898)

Serial Numbers :—
3313—3455
(less gaps)

Light passenger traffic engines.
(See note re 3200 class opposite.)

CYLINDERS—Diam., 18″; Stroke, 26″.
BOILER—Barrel, 11′ 0″; Diam. Outs., 4′ 5⅛″ and 5′ 0½″.
FIREBOX—Outs., 7′ 0″ by 5′ 3″ and 4′ 0″; Ins., 6′ 2 11/16″ by 4′ 3½″ and 3′ 2¾″; Height, 6′ 0 7/16″ and 5′ 0 7/16″.
TUBES—Superheater Tubes, No. 36: Diam., 1″: Length, 11′ 5⅜″. Fire Tubes, No. 218; Diam., 1⅝″; Length, 11′ 4 5/16″. Fire Tubes, No. 6: Diam., 5⅛″; Length, 11′ 4 5/16″.
TOTAL WEIGHT OF ENGINE—
51 tons 16 cwt. Full.
48 ,, 15 ,, Empty.

HEATING SURFACE—Superheater Tubes 82.30 sq. ft. Fire Tubes 1,145 sq. ft. Firebox 121.80 sq. ft. Total 1,349.10 sq. ft.
AREA OF FIREGRATE—20.35 sq. ft.
WHEELS—Bogie, 3′ 8″: Coupled, 5′ 8″.
WATER CAPACITY OF TENDER—3,500 gallons.
WORKING PRESSURE—200 lbs.
TRACTIVE EFFORT—21,060 lbs.
TOTAL WEIGHT OF TENDER—
40 tons 0 cwt. Full.
18 ,, 5 ,, Empty.

No.	Name.	No.	Name.	No.	Name.
3341	Blasius	3399	Ottawa	3443	Chaffinch
3353	Pershore Plum	3400	Winnipeg	3444	Cormorant
3363	Alfred Baldwin	3401	Vancouver	3445	Flamingo
3364	Frank Bibby	3406	Calcutta	3446	Goldfinch
3375	Sir Watkin Wynn	3407	Madras	3447	Jackdaw
3376	River Plym	3408	Bombay	3448	Kingfisher
3379	River Fal	3417	Lord Mildmay of Flete	3449	Nightingale
3391	Dominion of Canada			3450	Peacock
		3418	Sir Arthur Yorke	3451	Pelican
3393	Australia	3430	Inchcape	3452	Penguin
3395	Tasmania	3441	Blackbird	3453	Seagull
3396	Natal Colony	3442	Bullfinch	3454	Skylark
				3455	Starling

3200 Class. Type 4—4—0

(Introduced 1936)

Serial Numbers :—
3200—3219

These engines have been built from the 3300 ("Bulldog") class with smaller boilers, and replace engines of the older 3252 ("Duke") class which are being condemned. They are employed in working light passenger traffic over the late Cambrian Railway.

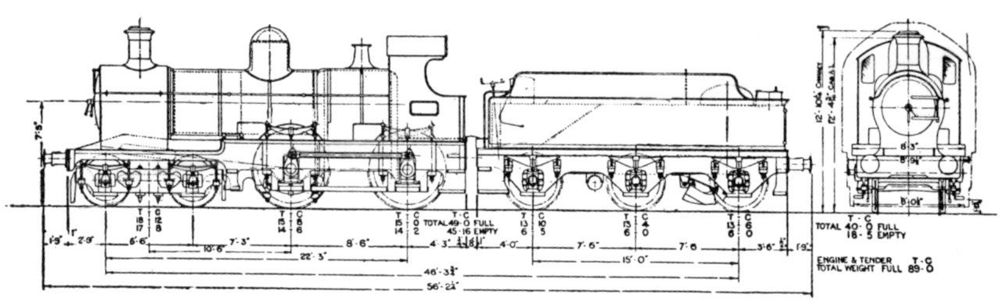

CYLINDERS—Diam., 18″ : Stroke, 26″

BOILER—Barrel, 11′ 0″ : Diam. Outs., 4′ 4″ and 4′ 5″.

FIREBOX—Outs., 5′ 10″ by 4′ 0″ ; Ins., 5′ 1 7/16″ by 3′ 4″ ; Height, 6′ 0$\frac{1}{2}$″.

TUBES—Superheater Tubes, No. 36 : Diam., 1″ ; Length, 11′ 1$\frac{3}{8}$″. Fire Tubes, No. 195 ; Diam., 1$\frac{5}{8}$″ ; Length, 11′ 3 11/16″. Fire Tubes, No. 6 : Diam., 5$\frac{1}{8}$″ ; Length, 11′ 3 11/16″.

TOTAL WEIGHT OF ENGINE—
49 tons 0 cwt. Full.
45 ,, 16 ,, Empty.

HEATING SURFACE—Superheater Tubes, 75.30 sq. ft. Fire Tubes 1,028.95 sq. ft. Firebox, 113.95 sq. ft. Total, 1,184.30 sq. ft.

AREA OF FIREGRATE—17.2 sq. ft.

WHEELS—Bogie, 3′ 8″ ; Coupled, 5′ 8″.

WATER CAPACITY OF TENDER—3,500 gallons.

WORKING PRESSURE—180 lbs.

TRACTIVE EFFORT—18,955 lbs.

TOTAL WEIGHT OF TENDER—
40 tons 0 cwt. Full.
18 ,, 5 ,, Empty.

Engines in 3200 Class are unnamed.

2900 Class. Type 4—6—0

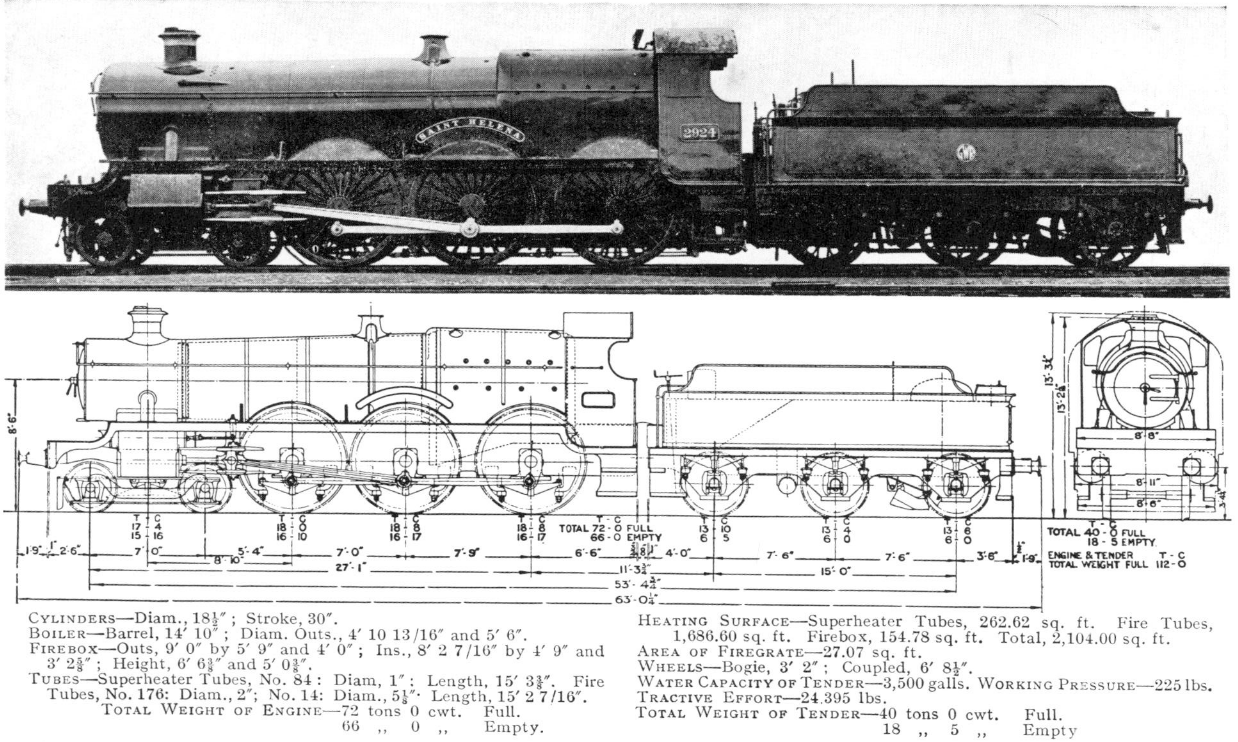

Cylinders—Diam., $18\frac{1}{2}''$; Stroke, 30″.
Boiler—Barrel, 14′ 10″; Diam. Outs., 4′ 10 13/16″ and 5′ 6″.
Firebox—Outs, 9′ 0″ by 5′ 9″ and 4′ 0″; Ins., 8′ 2 7/16″ by 4′ 9″ and 3′ $2\frac{5}{8}''$; Height, 6′ $6\frac{3}{8}''$ and 5′ $0\frac{3}{8}''$.
Tubes—Superheater Tubes, No. 84: Diam, 1″; Length, 15′ $3\frac{3}{8}''$. Fire Tubes, No. 176: Diam., 2″; No. 14: Diam., $5\frac{1}{8}''$. Length, 15′ 2 7/16″.
Total Weight of Engine—72 tons 0 cwt. Full.
66 ,, 0 ,, Empty.

Heating Surface—Superheater Tubes, 262.62 sq. ft. Fire Tubes, 1,686.60 sq. ft. Firebox, 154.78 sq. ft. Total, 2,104.00 sq. ft.
Area of Firegrate—27.07 sq. ft.
Wheels—Bogie, 3′ 2″; Coupled, 6′ $8\frac{1}{2}''$.
Water Capacity of Tender—3,500 galls. Working Pressure—225 lbs.
Tractive Effort—24.395 lbs.
Total Weight of Tender—40 tons 0 cwt. Full.
18 ,, 5 ,, Empty

"Saint" (2900) Class

(Introduced 1902)

Serial Numbers :—
2902—2989
(less gaps)

Express passenger traffic engines.

No.	Name.	No.	Name.
2902	Lady of the Lake	2940	Dorney Court
2903	Lady of Lyons	2941	Easton Court
2905	Lady Macbeth	2942	Fawley Court
2906	Lady of Lynn	2943	Hampton Court
2908	Lady of Quality	2944	Highnam Court
2912	Saint Ambrose	2945	Hillingdon Court
2913	Saint Andrew	2946	Langford Court
2915	Saint Bartholomew	2947	Madresfield Court
2916	Saint Benedict	2948	Stackpole Court
2920	Saint David	2949	Stanford Court
2924	Saint Helena	2950	Taplow Court
2926	Saint Nicholas	2951	Tawstock Court
2927	Saint Patrick	2952	Twineham Court
2928	Saint Sebastian	2953	Titley Court
2929	Saint Stephen	2954	Tockenham Court
2930	Saint Vincent	2955	Tortworth Court
2931	Arlington Court	2978	Charles J. Hambro
2932	Ashton Court	2979	Quentin Durward
2933	Bibury Court	2980	Cœur de Lion
2934	Butleigh Court	2981	Ivanhoe
2935	Caynham Court	2987	Bride of Lammermoor
2936	Cefntilla Court	2988	Rob Roy
2937	Clevedon Court	2989	Talisman
2938	Corsham Court		
2939	Croome Court		

4900 Class. Type 4—6—0

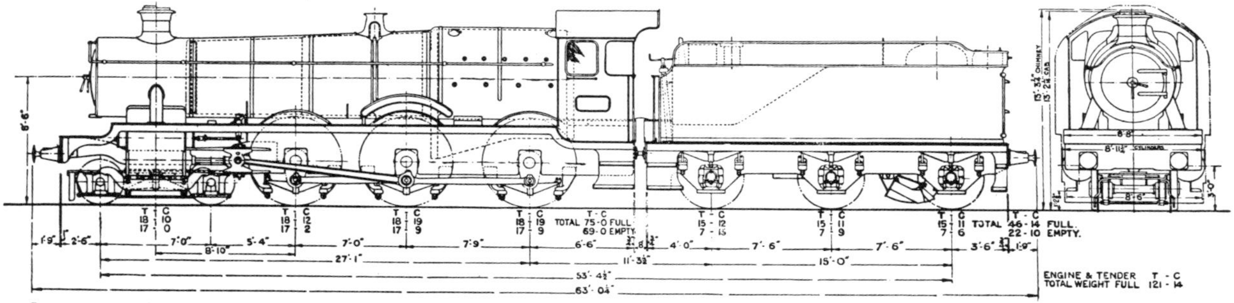

Cylinders—Diam., 18½″; Stroke, 30″.
Boiler—Barrel, 14′ 10″; Diam., Outs., 4′ 10 13/16″ and 5′ 6″.
Firebox—Outs., 9′ 0″ by 5′ 9″ and 4′ 0″. Ins., 8′ 2 7/16″ by 4′ 9″ and 3′ 2⅝″; Height, 6′ 6⅜″ and 5′ 0⅜″.
Tubes—Superheater Tubes, No. 84: Diam., 1″: Length, 15′ 3⅜″. Fire Tubes, No. 14: Diam., 5⅛″. No. 176: Diam., 2″; Length, 15′ 2 7/16″.
Total Weight of Engine—75 tons Full; 69 tons Empty.

Heating Surface—Superheater Tubes, 262.62 sq. ft. Fire Tubes, 1,686.60 sq. ft. Firebox, 154.78 sq. ft. Total, 2,104.0 sq. ft.
Area of Firegrate—27.07 sq. ft.
Wheels—Bogie, 3′ 0″; Coupled, 6′ 0″.
Water Capacity of Tender—4,000 gallons.
Working Pressure—225 lbs.
Tractive Effort—27,275 lbs.
Total Weight of Tender—46 tons 14 cwt. Full; 22 tons 10 cwt. Empty.

"Hall" (4900) Class

Serial Numbers :—
4900—4999
5900—5999
6900—6970

(Introduced 1928)

Mixed and excursion traffic engines.

No.	Name.	No.	Name.	No.	Name.
4900	Saint Martin	4947	Nanhoran Hall	4993	Dalton Hall
4901	Adderley Hall	4948	Northwick Hall	4994	Downton Hall
4902	Aldenham Hall	4949	Packwood Hall	4995	Easton Hall
4903	Astley Hall	4950	Patshull Hall	4996	Eden Hall
4904	Binnegar Hall	4951	Pendeford Hall	4997	Elton Hall
4905	Barton Hall	4952	Peplow Hall	4998	Eyton Hall
4906	Bradfield Hall	4953	Pitchford Hall	4999	Gopsal Hall
4907	Broughton Hall	4954	Plaish Hall	5900	Hinderton Hall
4908	Broome Hall	4955	Plaspower Hall	5901	Hazel Hall
4909	Blakesley Hall	4956	Plowden Hall	5902	Howick Hall
4910	Blaisdon Hall	4957	Postlip Hall	5903	Keele Hall
4912	Berrington Hall	4958	Priory Hall	5904	Kelham Hall
4913	Baglan Hall	4959	Purley Hall	5905	Knowsley Hall
4914	Cranmore Hall	4960	Pyle Hall	5906	Lawton Hall
4915	Condover Hall	4961	Pyrland Hall	5907	Marble Hall
4916	Crumlin Hall	4962	Ragley Hall	5908	Moreton Hall
4917	Crosswood Hall	4963	Rignall Hall	5909	Newton Hall
4918	Dartington Hall	4964	Rodwell Hall	5910	Park Hall
4919	Donnington Hall	4965	Rood Ashton Hall	5911	Preston Hall
4920	Dumbleton Hall	4966	Shakenhurst Hall	5912	Queen's Hall
4921	Eaton Hall	4967	Shirenewton Hall	5913	Rushton Hall
4922	Enville Hall	4968	Shotton Hall	5914	Ripon Hall
4923	Evenley Hall	4969	Shrugborough Hall	5915	Trentham Hall
4924	Eydon Hall	4970	Sketty Hall	5916	Trinity Hall
4925	Eynsham Hall	4971	Stanway Hall	5917	Westminster Hall
4926	Fairleigh Hall	4972	Saint Bride's Hall	5918	Walton Hall
4927	Farnborough Hall	4973	Sweeney Hall	5919	Worsley Hall
4928	Gatacre Hall	4974	Talgarth Hall	5920	Wycliffe Hall
4929	Goytrey Hall	4975	Umberslade Hall	5921	Bingley Hall
4930	Hagley Hall	4976	Warfield Hall	5922	Caxton Hall
4931	Hanbury Hall	4977	Watcombe Hall	5923	Colston Hall
4932	Hatherton Hall	4978	Westwood Hall	5924	Dinton Hall
4933	Himley Hall	4979	Wootton Hall	5925	Eastcote Hall
4934	Hindlip Hall	4980	Wrottesley Hall	5926	Grotrian Hall
4935	Ketley Hall	4981	Abberley Hall	5927	Guild Hall
4936	Kinlet Hall	4982	Acton Hall	5928	Haddon Hall
4937	Lanelay Hall	4983	Albert Hall	5929	Hanham Hall
4938	Liddington Hall	4984	Albrighton Hall	5930	Hannington Hall
4939	Littleton Hall	4985	Allesley Hall	5931	Hatherley Hall
4940	Ludford Hall	4986	Aston Hall	5932	Haydon Hall
4941	Llangedwyn Hall	4987	Brockley Hall	5933	Kingsway Hall
4942	Maindy Hall	4988	Bulwell Hall	5934	Kneller Hall
4943	Marrington Hall	4989	Cherwell Hall	5935	Norton Hall
4944	Middleton Hall	4990	Clifton Hall	5936	Oakley Hall
4945	Milligan Hall	4991	Cobham Hall	5937	Stanford Hall
4946	Moseley Hall	4992	Crosby Hall	5938	Stanley Hall

"Hall" (4900) Class—continued.

No.	Name.	No.	Name.	No.	Name.
5939	Tangley Hall	5983	Henley Hall	6927	Lilford Hall
5940	Whitbourne Hall	5984	Linden Hall	6928	Underley Hall
5941	Campion Hall	5985	Mostyn Hall	6929	Whorlton Hall
5942	Doldowlod Hall	5986	Arbury Hall	6930	Aldersey Hall
5943	Elmdon Hall	5987	Brocket Hall	6931	Aldborough Hall
5944	Ickenham Hall	5988	Bostock Hall	6932	Burwarton Hall
5945	Leckhampton Hall	5989	Cransley Hall	6933	Birtles Hall
5946	Marwell Hall	5990	Dorford Hall	6934	Beachamwell Hall
5947	Saint Benet's Hall	5991	Gresham Hall	6935	Browsholme Hall
5948	Siddington Hall	5992	Horton Hall	6936	Breccles Hall
5949	Trematon Hall	5993	Kirby Hall	6937	Conyngham Hall
5950	Wardley Hall	5994	Roydon Hall	6938	Corndean Hall
5951	Clyffe Hall	5995	Wick Hall	6939	Calveley Hall
5952	Cogan Hall	5996	Mytton Hall	6940	Didlington Hall
5953	Dunley Hall	5997	Sparkford Hall	6941	Fillongley Hall
5954	Faendre Hall	5998	Trevor Hall	6942	Eshton Hall
5955	Garth Hall	5999	Wollaton Hall	6943	Farnley Hall
5956	Horsley Hall	6900	Abney Hall	6944	Fledborough Hall
5957	Hutton Hall	6901	Arley Hall	6945	Glasfryn Hall
5958	Knolton Hall	6902	Butlers Hall	6946	Heatherden Hall
5959	Mawley Hall	6903	Belmont Hall	6947	Helmingham Hall
5960	Saint Edmund Hall	6904	Charfield Hall	6948	Holbrooke Hall
5961	Toynbee Hall	6905	Claughton Hall	6949	Haberfield Hall
5962	Wantage Hall	6906	Chicheley Hall	6950	Kingsthorpe Hall
5963	Wimpole Hall	6907	Davenham Hall	6951	Impney Hall
5964	Wolseley Hall	6908	Downham Hall	6952	Kimberley Hall
5965	Woollas Hall	6909	Frewin Hall	6953	Leighton Hall
5966	Ashford Hall	6910	Gossington Hall	6954	Lotherton Hall
5967	Bickmarsh Hall	6911	Holker Hall	6955	Lydcott Hall
5968	Cory Hall	6912	Helmster Hall	6956	Mottram Hall
5969	Honington Hall	6913	Levens Hall	6957	Norcliffe Hall
5970	Hengrave Hall	6914	Langton Hall	6958	Oxburgh Hall
5971	Merevale Hall	6915	Mursley Hall	6959	Peatling Hall
5972	Olton Hall	6916	Misterton Hall	6960	Raveningham Hall
5973	Rolleston Hall	6917	Oldlands Hall	6961	Stedham Hall
5974	Wallsworth Hall	6918	Sandon Hall	6962	Soughton Hall
5975	Winslow Hall	6919	Tylney Hall	6963	Throwley Hall
5976	Ashwicke Hall	6920	Barningham Hall	6964	Thornbridge Hall
5977	Beckford Hall	6921	Borwick Hall	6965	Thirlestaine Hall
5978	Bodinnick Hall	6922	Burton Hall	6966	Witchingham Hall
5979	Cruckton Hall	6923	Croxteth Hall	6967	Willesley Hall
5980	Dingley Hall	6924	Grantley Hall	6968	Woodcock Hall
5981	Frensham Hall	6925	Hackness Hall	6969	Wraysbury Hall
5982	Harrington Hall	6926	Holkham Hall	6970	Whaddon Hall

A number of these engines are listed for conversion to oil burning. These will eventually be renumbered under a new 3900 group.

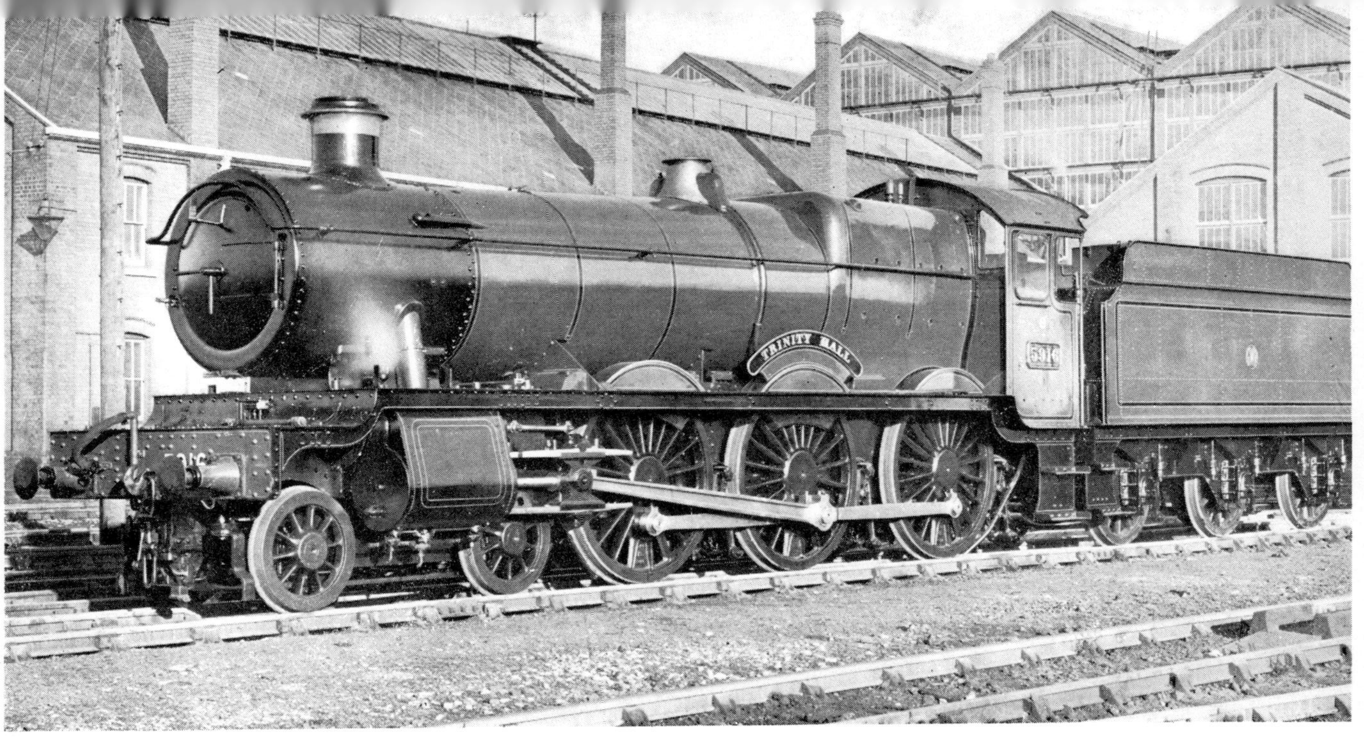

Locomotive No. 5916 "Trinity Hall."

6800 Class. Type 4—6—0

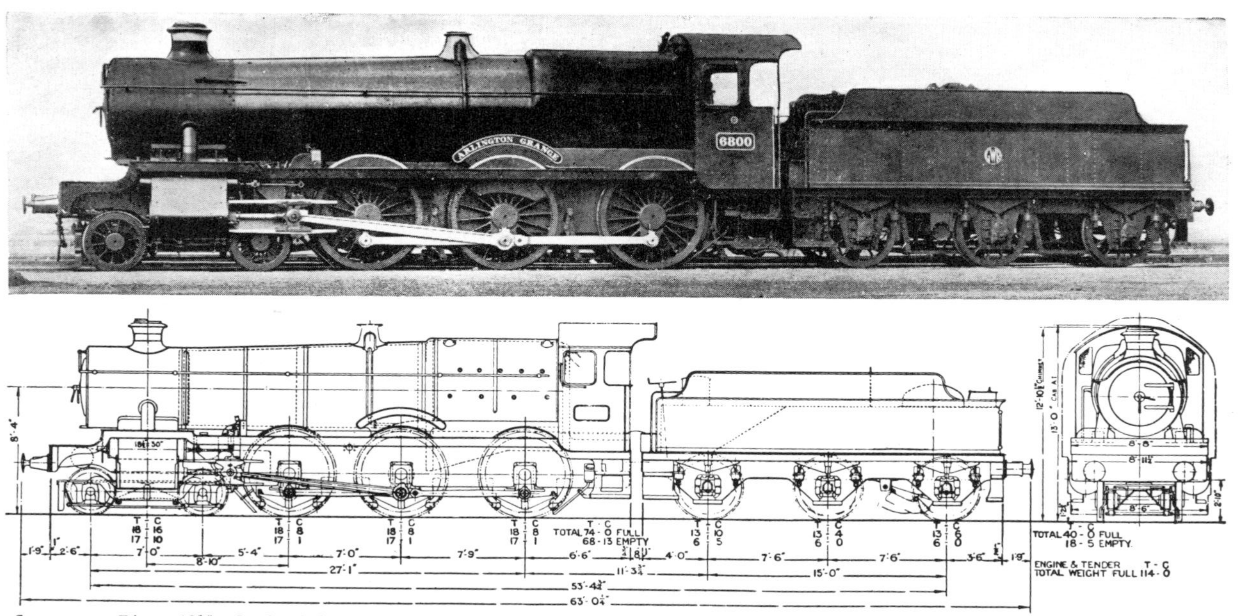

CYLINDERS—Diam., 18½″; Stroke, 30″.
BOILER—Barrel, 14′ 10″; Diam., Outs., 4′ 10 13/16″ and 5′ 6″.
FIREBOX—Outs., 9′ 0″ by 5′ 9″ and 4′ 0″. Ins., 8′ 2 7/16″ by 4′ 9″ and 3′ 2⅝″; Height, 6′ 6⅜″ and 5′ 0⅜″.
TUBES—Superheater Tubes, No. 84: Diam., 1″; Length, 15′ 3⅜″. Fire Tubes, No. 14: Diam. 5⅛″., No. 176: Diam., 2″; Length, 15′ 2 7/16″.

HEATING SURFACE—Superheater Tubes, 262.62 sq. ft. Fire Tubes, 1,686.60 sq. ft. Firebox, 154. 78 sq. ft. Total, 2,104.0 sq. ft.
AREA OF FIREGRATE—27.07 sq. ft.
WHEELS—Bogie 3′ 0″. Coupled, 5′ 8″.
WATER CAPACITY, OF TENDER—3,500 gallons.
WORKING PRESSURE—225 lbs. TRACTIVE EFFORT—28,875 lbs.

"Grange" (6800) Class
(Introduced 1936)

**Serial Numbers :—
6800—6879**

Designed to replace older 4300 class 2—6—0 type for working mixed traffic.

No.	Name.	No.	Name.
6800	Arlington Grange	6840	Hazeley Grange
6801	Aylburton Grange	6841	Marlas Grange
6802	Bampton Grange	6842	Nunhold Grange
6803	Bucklebury Grange	6843	Poulton Grange
6804	Brockington Grange	6844	Penhydd Grange
6805	Broughton Grange	6845	Paviland Grange
6806	Blackwell Grange	6846	Ruckley Grange
6807	Birchwood Grange	6847	Tidmarsh Grange
6808	Beenham Grange	6848	Toddington Grange
6809	Burghclere Grange	6849	Walton Grange
6810	Blakemere Grange	6850	Cleeve Grange
6811	Cranbourne Grange	6851	Hurst Grange
6812	Chesford Grange	6852	Headbourne Grange
6813	Eastbury Grange	6853	Morehampton Grange
6814	Enborne Grange	6854	Roundhill Grange
6815	Frilford Grange	6855	Saighton Grange
6816	Frankton Grange	6856	Stowe Grange
6817	Gwenddwr Grange	6857	Tudor Grange
6818	Hardwick Grange	6858	Woolston Grange
6819	Highnam Grange	6859	Yiewsley Grange
6820	Kingstone Grange	6860	Aberporth Grange
6821	Leaton Grange	6861	Crynant Grange
6822	Manton Grange	6862	Derwent Grange
6823	Oakley Grange	6863	Dolhywel Grange
6824	Ashley Grange	6864	Dymock Grange
6825	Llanvair Grange	6865	Hopton Grange
6826	Nannerth Grange	6866	Morfa Grange
6827	Llanfrechfa Grange	6867	Peterston Grange
6828	Trellech Grange	6868	Penrhos Grange
6829	Burmington Grange	6869	Resolven Grange
6830	Buckenhill Grange	6870	Bodicote Grange
6831	Bearley Grange	6871	Bourton Grange
6832	Brockton Grange	6872	Crawley Grange
6833	Calcot Grange	6873	Caradoc Grange
6834	Dummer Grange	6874	Haughton Grange
6835	Eastham Grange	6875	Hindford Grange
6836	Estevarney Grange	6876	Kingsland Grange
6837	Forthampton Grange	6877	Llanfair Grange
6838	Goodmoor Grange	6878	Longford Grange
6839	Hewell Grange	6879	Overton Grange

7800 Class. Type 4—6—0

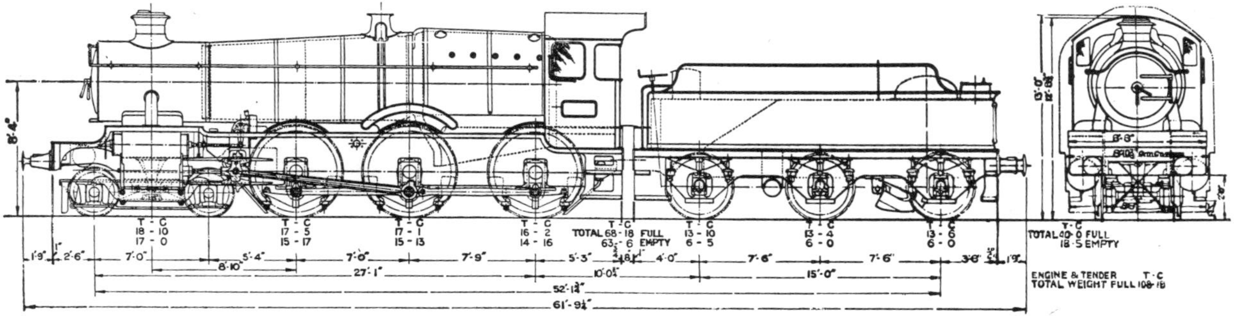

CYLINDERS—Diam., 18″; Stroke, 30″.
BOILER—Barrel, 12′6″; Diam. Outs., 4′ 7⅝″ and 5′ 3″.
FIREBOX—Outs., 8′8⅛″ by 5′ 5½″ and 7′ 6″ by 4′ 0″; Ins., 7′ 8 5/16 by 4′ 5¾″ and 6′ 8 11/16″ by 3′ 2¾″; Height, 6′ 3 7/16″ and 4′ 10 7/16″.
TUBES—Superheater Tubes, No. 72: Diam., 1″; Length, 13′ 5 13/16″. Fire Tubes, No. 12: Diam., 5⅛″; No. 158: Diam., 2″; Length 13′ 0 5/16″.

HEATING SURFACE—Superheater Tubes, 160.0 sq. ft. Fire Tubes, 1,285.5 sq. ft. Firebox, 140.0 sq. ft. Total, 1,585.5 sq. ft.
AREA OF FIREGRATE—22.1 sq. ft.
WHEELS—Bogie, 3′ 0″; Coupled, 5′ 8″.
WATER CAPACITY OF TENDER—3,500 gallons
WORKING PRESSURE—225 lbs.
TRACTIVE EFFORT—27,340 lbs.

"Manor" (7800) Class

(Introduced 1938)

Serial Numbers :—
7800—7819

Designed to replace older 4300 class 2—6—0 type, for working mixed traffic.

No.	Name.	No.	Name.
7800	Torquay Manor		
7801	Anthony Manor		
7802	Bradley Manor		
7803	Barcote Manor		
7804	Baydon Manor		
7805	Broome Manor		
7806	Cockington Manor		
7807	Compton Manor		
7808	Cookham Manor		
7809	Childrey Manor		
7810	Draycott Manor		
7811	Dunley Manor		
7812	Erlestoke Manor		
7813	Freshford Manor		
7814	Fringford Manor		
7815	Fritwell Manor		
7816	Frilsham Manor		
7817	Garsington Manor		
7818	Granville Manor		
7819	Hinton Manor		

1000 Class. Type 4—6—0

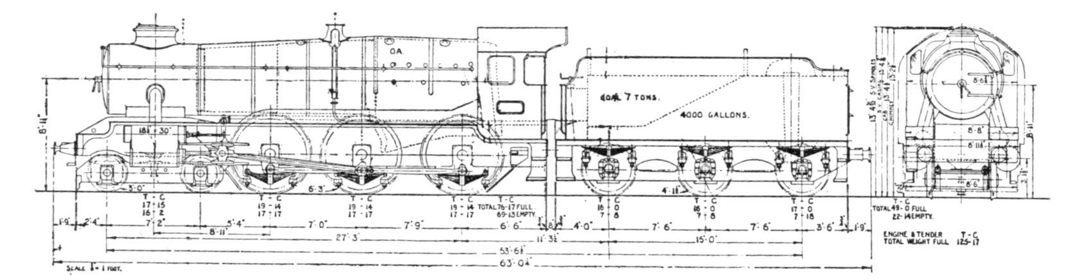

CYLINDERS—Diam., 18½″; Stroke, 30″.
BOILER—Barrel, 12′ 7 3/16″; Diam. Outs., 5′ 0″ and 5′ 8⅜″.
FIREBOX—Outs., 9′ 9″ by 5′ 10⅞″ and 9′ 3″ by 4′ 0″; Ins., 8′ 7 13/16″ by 4′ 9½″ and 8′ 6 7/16″ by 3′ 3⅝″; Height, 6′ 8 11/16″ and 5′ 1 23/32″.
TUBES—Superheater Tubes, No. 84: Diam., 1¼″; Length, 11′ 6″. Fire Tubes, No. 21: Diam., 5⅛″, No. 198, Diam, 1¾″. Length, 13′ 0″.
TOTAL WEIGHT OF ENGINE—76 tons 17 cwt. Full.
69 ,, 13 ,, Empty.

HEATING SURFACE—Superheater Tubes, 254 sq. ft. Fire Tubes, 1,545 sq. ft. Firebox, 169 sq. ft. Total, 1,968 sq. ft.
AREA OF FIREGRATE—28.84 sq. ft.
WHEELS—Bogie, 3′ 0″: Coupled, 6′ 3″.
WATER CAPACITY OF TENDER—4,000 gallons.
WORKING PRESSURE—280 lbs. per sq. in.
TRACTIVE EFFORT—32,580 lbs.
TOTAL WEIGHT OF TENDER—49 tons. Full.
22 tons, 14 cwt. Empty.

"County" (1000) Class

(Introduced 1945)

Serial Numbers :—
1000—1029 *

Mixed traffic engines.

No.	Name.	No.	Name.
1000	County of Middlesex		
1001	County of Bucks		
1002	County of Berks		
1003	County of Wilts		
1004	County of Somerset		
1005	County of Devon		
1006	County of Cornwall		
1007	County of Brecknock		
1008	County of Cardigan		
1009	County of Carmarthen		
1010	County of Carnarvon		
1011	County of Chester		
1012	County of Denbigh		
1013	County of Dorset		
1014	County of Glamorgan		
1015	County of Gloucester		
1016	County of Hants		
1017	County of Hereford		
1018	County of Leicester		
1019	County of Merioneth		
1020	County of Monmouth		
1021	County of Montgomery		
1022	County of Northampton		
1023	County of Oxford		
1024	County of Pembroke		
1025	County of Radnor		
1026	County of Salop		
1027	County of Stafford		
1028	County of Warwick		
1029	County of Worcester		

*Engines in service, 1000—1019. Engines now under construction in series, 1020—1029

4000 Class. Type 4—6—0

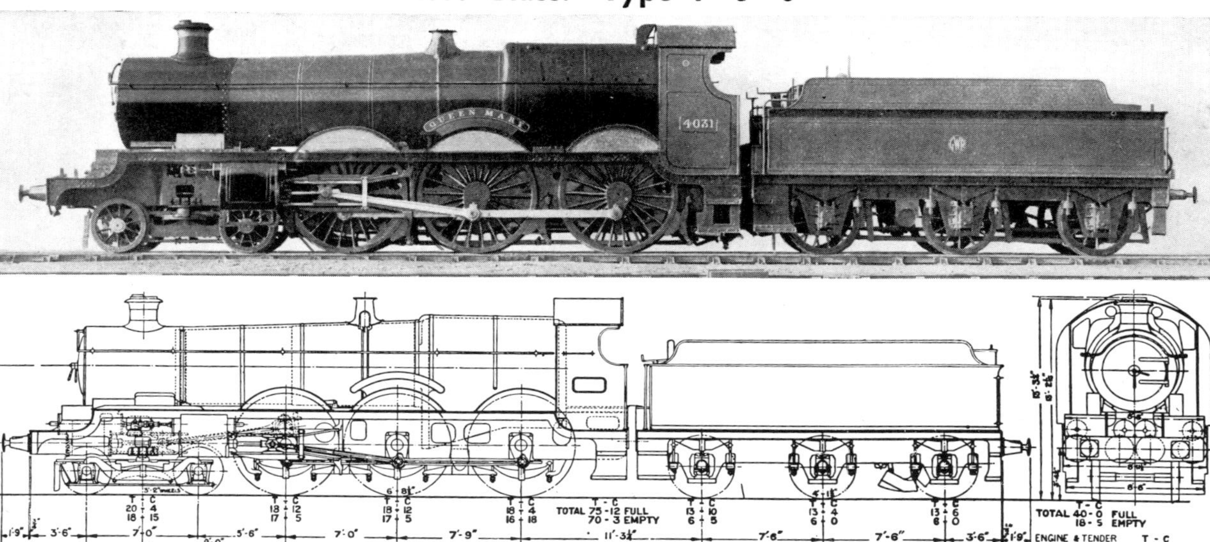

CYLINDERS—Four: Diam. 15″; Stroke, 26″.
BOILER—Barrel, 14′ 10″; Diam., Outs., 4′ 10 13/16″ and 5′ 6″.
FIREBOX—Outs., 9′ 0″ by 5′ 9″ and 4′ 0″; Ins., 8′ 2 7/16″ by 4′ 9″ and 3′ 2⅝″; Height, 6′ 6⅜″ and 5′ 0⅜″.
TUBES—Superheater Tubes, No. 84: Diam., 1″; Length, 15′ 3⅜″. Fire Tubes, No. 176: Diam., 2″; Length, 15′ 2 7/16″. Fire Tubes, No. 14: Diam., 5⅛″; Length, 15′ 2 7/16″

HEATING SURFACE—Superheater Tubes, 262.62 sq. ft. Fire Tubes, 1,686.60 sq. ft. Firebox, 154.78 sq. ft. Total, 2,104.0 sq. ft.
AREA OF FIREGRATE—27.07 sq. ft. WHEELS—Bogie, 3′ 2″; Coupled, 6′ 8½″.
WATER CAPACITY OF TENDER—4,000 gallons.
WORKING PRESSURE—225 lbs. TRACTIVE EFFORT—27,800 lbs.
TOTAL WEIGHT OF ENGINE—75 tons 12 cwt. Full; 70 tons 3 cwt. Empty.
TOTAL WEIGHT OF TENDER—46 tons 14 cwt. Full; 22 tons 10 cwt. Empty.

"Star" (4000) Class. Four-cylinder 4—6—0 Type
(Introduced 1906)

Serial Numbers :—
4003—4062
(less gaps)

The first engine in this class (then No. 40) was built with a 4—4—2 wheel arrangement experimentally in 1906 and later (1909) rebuilt as 4—6—0 type. These engines are employed on main line express passenger services.

No.	Name.	No.	Name.
4003	Lode Star	4040	Queen Boadicea
4004	Morning Star	4041	Prince of Wales
4007	Swallowfield Park	4042	Prince Albert
4012	Knight of the Thistle	4043	Prince Henry
4013	Knight of St. Patrick	4044	Prince George
4015	Knight of St. John	4045	Prince John
4017	Knight of Liège	4046	Princess Mary
4018	Knight of the Grand Cross	4047	Princess Louise
4019	Knight Templar	4048	Princess Victoria
4020	Knight Commander	4049	Princess Maud
4021	British Monarch	4050	Princess Alice
4022*		4051	Princess Helena
4023*		4052	Princess Beatrice
4025*		4053	Princess Alexandra
4026*		4054	Princess Charlotte
4028*		4055	Princess Sophia
4030*		4056	Princess Margaret
4031	Queen Mary	4057	Princess Elizabeth
4033	Queen Victoria	4058	Princess Augusta
4034	Queen Adelaide	4059	Princess Patricia
4035	Queen Charlotte	4060	Princess Eugenie
4036	Queen Elizabeth	4061	Glastonbury Abbey
4038	Queen Berengaria	4062	Malmesbury Abbey
4039	Queen Matilda		

* Nameplates removed.

4073 Class. Type 4—6—0

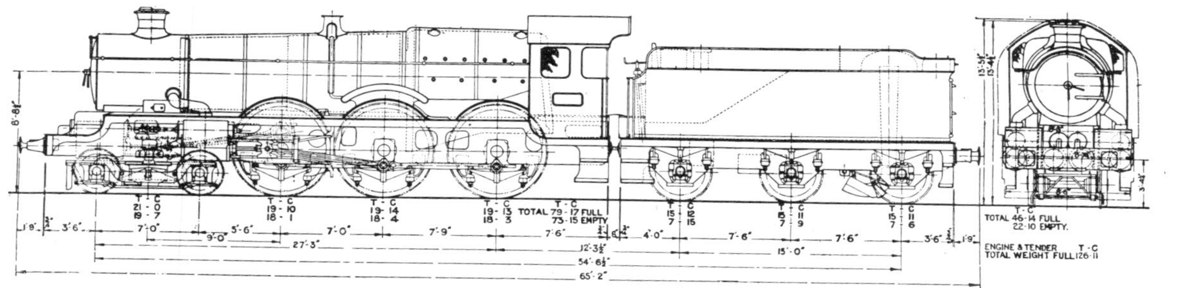

Cylinders—Four: Diam., 16″; Stroke, 26″.
Boiler—Barrel, 14′ 10″; Diam. Outs., 5′ 1 15/16″ and 5′ 9″.
Firebox—Outs., 10′ 0″ by 6′ 0″ and 4′ 0″; Ins., 9′ 2 7/16″ by 5′ 0⅛″ and 3′ 2⅝″; Height, 6′ 8⅞″ and 5′ 3⅞″.
Tubes—Superheater Tubes, No. 84: Diam., 1″; Length, 15′ 3⅜″. Fire Tubes, No. 201: Diam., 2″; No. 14: Diam., 5⅛″; Length, 15′ 2 7/16″.

Total Weight of Engine—79 tons 17 cwt. Full.
73 ,, 15 ,, Empty.

Heating Surface—Superheater Tubes, 262.62 sq. ft. Fire Tubes, 1,885.62 sq. ft. Firebox, 163.76 sq. ft. Total, 2,312.0 sq. ft.
Area of Firegrate—29.36 sq. ft.
Wheels—Bogie, 3′ 2″; Coupled, 6′ 8½″.
Water Capacity of Tender—4,000 gallons.
Working Pressure—225 lbs.
Tractive Effort—31,625 lbs.
Total Weight of Tender—46 tons 14 cwt. Full.
22 ,, 10 ,, Empty.

"Castle" (4073) Class

Four-cylinder 4—6—0 Type (Introduced 1923)

Serial Numbers :—

4073—4099
5000—5099
7000—7007

These engines have larger boiler and cylinders than the earlier "Star" (4000) class. Designed for main line express passenger services.

No.	Name.	No.	Name.	No.	Name.
100.A1	Lloyd's	5013	Abergavenny Castle	5061	Earl of Birkenhead
111	Viscount Churchill	5014	Goodrich Castle	5062	Earl of Shaftesbury
4000	North Star	5015	Kingswear Castle	5063	Earl Baldwin
4016	The Somerset Light	5016	Montgomery Castle	5064	Bishops' Castle
	Infantry (Prince	5017	St. Donats Castle	5065	Newport Castle
	Albert's)	5018	St. Mawes Castle	5066	Wardour Castle
4032	Queen Alexandra	5019	Treago Castle	5067	St. Fagans Castle
4037	The South Wales	5020	Trematon Castle	5068	Beverston Castle
	Borderers	5021	Whittington Castle	5069	Isambard Kingdom
4073	Caerphilly Castle	5022	Wigmore Castle		Brunel
4074	Caldicot Castle	5023	Brecon Castle	5070	Sir Daniel Gooch
4075	Cardiff Castle	5024	Carew Castle	5071	Spitfire
4076	Carmarthen Castle	5025	Chirk Castle	5072	Hurricane
4077	Chepstow Castle	5026	Criccieth Castle	5073	Blenheim
4078	Pembroke Castle	5027	Farleigh Castle	5074	Hampden
4079	Pendennis Castle	5028	Llantilio Castle	5075	Wellington
4080	Powderham Castle	5029	Nunney Castle	5076	Gladiator
4081	Warwick Castle	5030	Shirburn Castle	5077	Fairey Battle
4082	Windsor Castle	5031	Totnes Castle	5078	Beaufort
4083	Abbotsbury Castle	5032	Usk Castle	5079	Lysander
4084	Aberystwyth Castle	5033	Broughton Castle	5080	Defiant
4085	Berkeley Castle	5034	Corfe Castle	5081	Lockheed Hudson
4086	Builth Castle	5035	Coity Castle	5082	Swordfish
4087	Cardigan Castle	5036	Lyonshall Castle	5083	Bath Abbey
4088	Dartmouth Castle	5037	Monmouth Castle	5084	Reading Abbey
4089	Donnington Castle	5038	Morlais Castle	5085	Evesham Abbey
4090	Dorchester Castle	5039	Rhuddlan Castle	5086	Viscount Horne
4091	Dudley Castle	5040	Stokesay Castle	5087	Tintern Abbey
4092	Dunraven Castle	5041	Tiverton Castle	5088	Llanthony Abbey
4093	Dunster Castle	5042	Winchester Castle	5089	Westminster Abbey
4094	Dynevor Castle	5043	Earl of Mount	5090	Neath Abbey
4095	Harlech Castle		Edgcumbe	5091	Cleeve Abbey
4096	Highclere Castle	5044	Earl of Dunraven	5092	Tresco Abbey
4097	Kenilworth Castle	5045	Earl of Dudley	5093	Upton Castle
4098	Kidwelly Castle	5046	Earl of Cawdor	5094	Tretower Castle
4099	Kilgerran Castle	5047	Earl of Dartmouth	5095	Barbury Castle
5000	Launceston Castle	5048	Earl of Devon	5096	Bridgwater Castle
5001	Llandovery Castle	5049	Earl of Plymouth	5097	Sarum Castle
5002	Ludlow Castle	5050	Earl of St. Germans	5098	Clifford Castle
5003	Lulworth Castle	5051	Earl Bathurst	5099	Compton Castle
5004	Llanstephan Castle	5052	Earl of Radnor	7000	Viscount Portal
5005	Manorbier Castle	5053	Earl Cairns	7001	Denbigh Castle
5006	Tregenna Castle	5054	Earl of Ducie	7002	Devizes Castle
5007	Rougemont Castle	5055	Earl of Eldon	7003	Elmley Castle
5008	Raglan Castle	5056	Earl of Powis	7004	Eastnor Castle
5009	Shrewsbury Castle	5057	Earl of Waldegrave	7005	Lamphey Castle
5010	Restormel Castle	5058	Earl of Clancarty	7006	Lydford Castle
5011	Tintagel Castle	5059	Earl St. Aldwyn	7007	Ogmore Castle
5012	BerryPomeroyCastle	5060	Earl of Berkeley		

6000 Class.

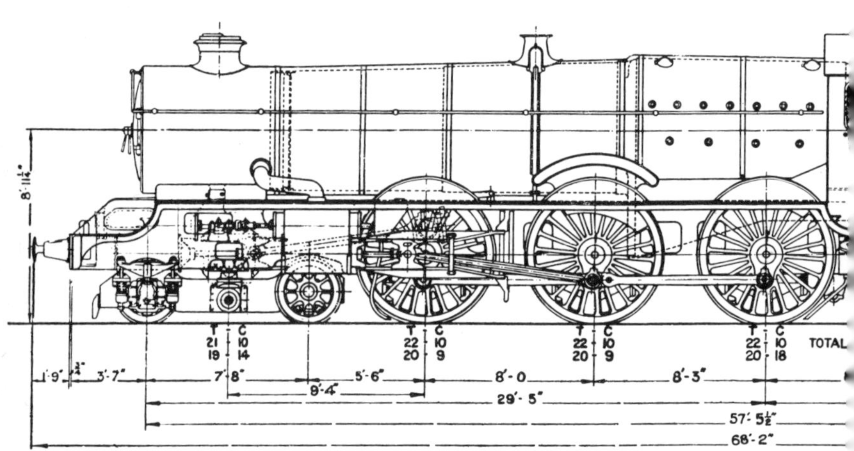

CYLINDERS—Four: Diam., 16¼″; Stroke, 28″.
BOILER—Barrel, 16′ 0″; Diam. Outs., 5′ 6¼″ and 6′ 0″.
FIREBOX—Length, Outs., 11′ 6″.

TOTAL WEIGHT OF ENGINE—89 tons 0 cwt. Full.
81 ,, 10 ,, Empty.

pe 4—6—0

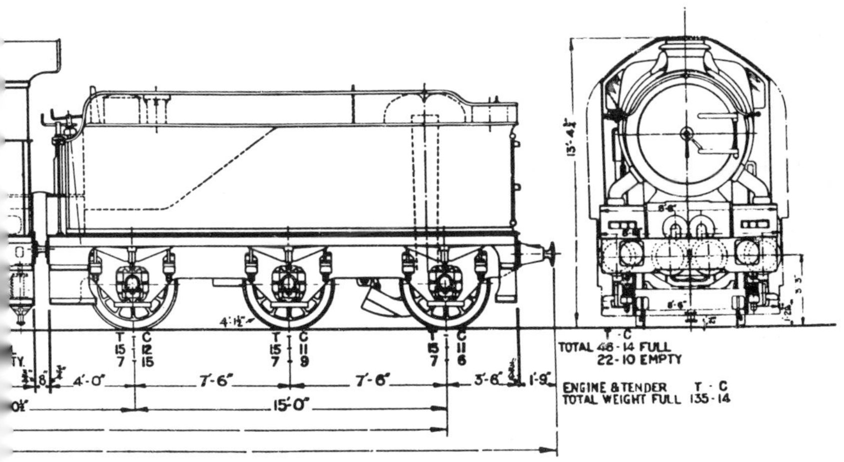

HEATING SURFACE—2,490 sq. ft.
AREA OF FIREGRATE—34.3 sq. ft.
WHEELS—Bogie, 3′ 0″ : Coupled, 6′ 6″
WATER CAPACITY OF TENDER—4,000 gallons.
WORKING PRESSURE—250 lbs.
TRACTIVE EFFORT—40,300 lbs.
TOTAL WEIGHT OF TENDER—46 tons 14 cwt. Full.
22 ,, 10 ,, Empty.

"King" (6000) Class
Four-cylinder 4—6—0 Type
(Introduced 1927)

These are the largest heavy express passenger locomotives on the Great Western Railway, and were specially designed to cope with heavy traffic at high speeds.

Serial Numbers 6000—6029

No.	Name.	No.	Name.
6000	King George V	6015	King Richard III
6001	King Edward VII	6016	King Edward V
6002	King William IV	6017	King Edward IV
6003	King George IV	6018	King Henry VI
6004	King George III	6019	King Henry V
6005	King George II	6020	King Henry IV
6006	King George I	6021	King Richard II
6007	King William III	6022	King Edward III
6008	King James II	6023	King Edward II
6009	King Charles II	6024	King Edward I
6010	King Charles I	6025	King Henry III
6011	King James I	6026	King John
6012	King Edward VI	6027	King Richard I
6013	King Henry VIII	6028	King George VI
6014	King Henry VII	6029	King Edward VIII

"King George V" was sent to America in 1927 together with the old (rebuilt) "North Star," representing G.W.R. locomotives ancient and modern, and was on view at the Baltimore and Ohio Railway Centenary Exhibition.

To commemorate the occasion, "King George V" was presented with the large brass bell shown on the opposite page, similar to those carried on American locomotives, and the medal here shown (in obverse and reverse).

The bell is now carried on the buffer plate of "King George V" and is inscribed as under :—

"Presented to Locomotive King George V
by the
Baltimore and Ohio Railroad Company
in commemoration of its
Centenary Celebration.
Sept. 24th—Oct. 15th, 1927."

The medal is mounted on the side of the engine cab over the number plate.

Locomotive No. 6000. "King George V" with the bell presented by the Baltimore and Ohio Railroad Company at their Centenary Celebrations, 1927

2600 Class. Type 2—6—0 (Introduced 1900) Serial Numbers :—2600—2680 (less gaps)

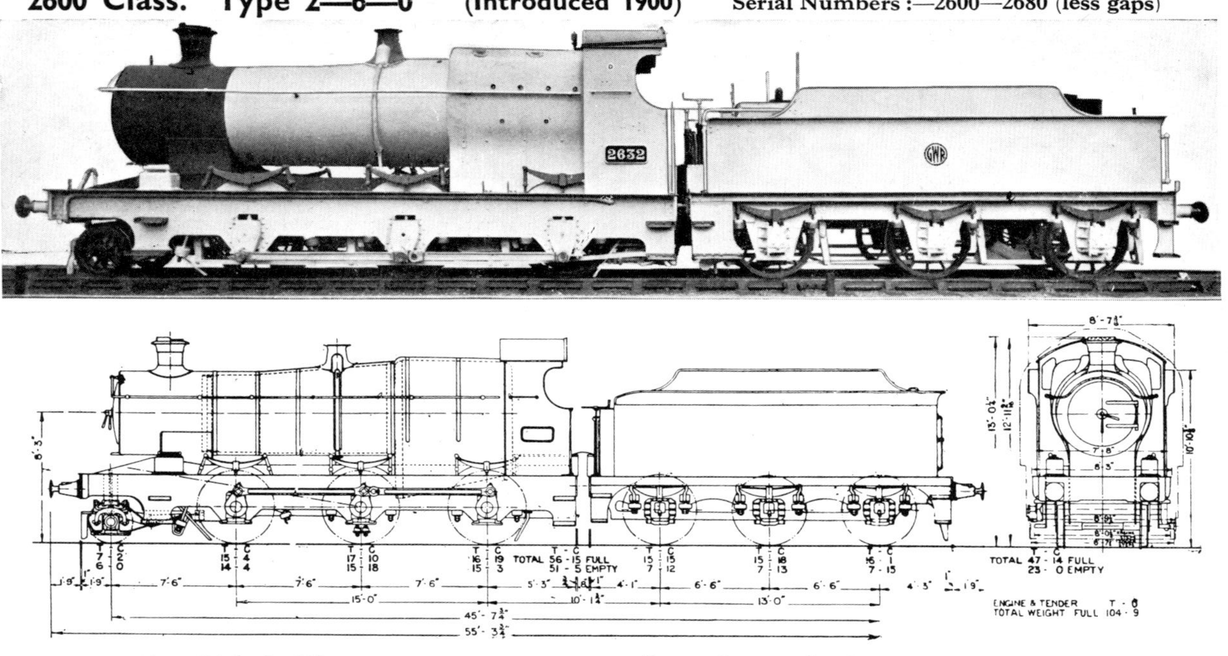

CYLINDERS—Diam., 18″ Stroke, 26″.
BOILER—Barrel, 11′ 0″; Diam. Outs., 4′ 10¾″ and 5′ 6″.
FIREBOX—Outs., 7′ 6″ by 5′ 9″ and 4′ 0″; Ins., 6′ 2½″ by 4′ 8⅞″ and 3′ 2⅝″; Height, 6′ 6⅜″ and 5′ 0⅜″.
TUBES—Superheater Tubes, No. 84: Diam., 1″; Length, 11′ 5⅜″. Fire Tubes, No. 235: Diam., 1⅝″; Length, 11′ 4 7/16″. Fire Tubes, No. 14: Diam., 5⅛″; Length, 11′ 4 7/16″.
TOTAL WEIGHT OF ENGINE—56 tons 15 cwt. Full.
51 ,, 5 ,, Empty.

HEATING SURFACE—Superheater Tubes, 191.88 sq. ft. Fire Tubes, 1,349.64 sq. ft. Firebox, 128.72 sq. ft. Total, 1,670.24 sq. ft.
AREA OF FIREGRATE—20.56 sq. ft.
WHEELS—Pony, 2′ 8″; Coupled, 4′ 7½″.
WATER CAPACITY OF TENDER—4,000 gallons
WORKING PRESSURE—200 lbs.
TRACTIVE EFFORT—25,800 lbs.
TOTAL WEIGHT OF TENDER—47 tons 14 cwt. Full.
23 ,, 0 ,, Empty.

These engines, also known as the "Aberdare" Class, were designed for working coal traffic in South Wales.

Serial Numbers :— 4300—4399 5300—5399 6300—6399 7300—7321 9300—9319

4300 Class. Type 2—6—0
(Introduced 1911) Mixed Traffic Engines

Note :—9300 class have modified weight distribution.

CYLINDERS—Diam., 18½″; Stroke, 30″.
BOILER—Barrel, 11′ 0″; Diam. Outs., 4′ 10¾″ and 5′ 6″.
FIREBOX—Outs., 7′ 0″ by 5′ 9″ and 4′ 0″; Ins., 6′ 2½″ by 4′ 8⅞″ and 3′ 2⅝″. Height, 6′ 6⅜″ and 5′ 0⅜″.
TUBES—Superheater Tubes, No. 84: Diam., 1″; Length, 11′ 5⅜″. Fire Tubes, No. 235: Diam., 1⅝″; Length, 11′ 4 7/16″. Fire Tubes, No. 14: Diam., 5⅛″; Length, 11′ 4 7/16″.
HEATING SURFACE—Superheater Tubes 191.79 sq. ft. Fire Tubes, 1,349.64 sq. ft. Firebox, 128.72 sq. ft. Total, 1,670.15 sq. ft.
AREA OF FIREGRATE—20.56 sq. ft.
WHEELS—Pony, 3′ 2″; Coupled, 5′ 8″.
WATER CAPACITY OF TENDER—3,500 gallons.
WORKING PRESSURE—200 lbs. TRACTIVE EFFORT—25,670 lbs.
TOTAL WEIGHT OF TENDER—40 tons 0 cwt. Full; 18 tons 5 cwt. Empty.]
(4300 Class). TOTAL WEIGHT OF ENGINE—62 tons 0 cwt. Full; 57 tons 14 cwt. Empty.

Serial Numbers :—
2800—2883 2884—2899
3800—3866

2800 Class. Type 2—8—0
(Introduced 1903)

Heavy Goods Traffic Engines

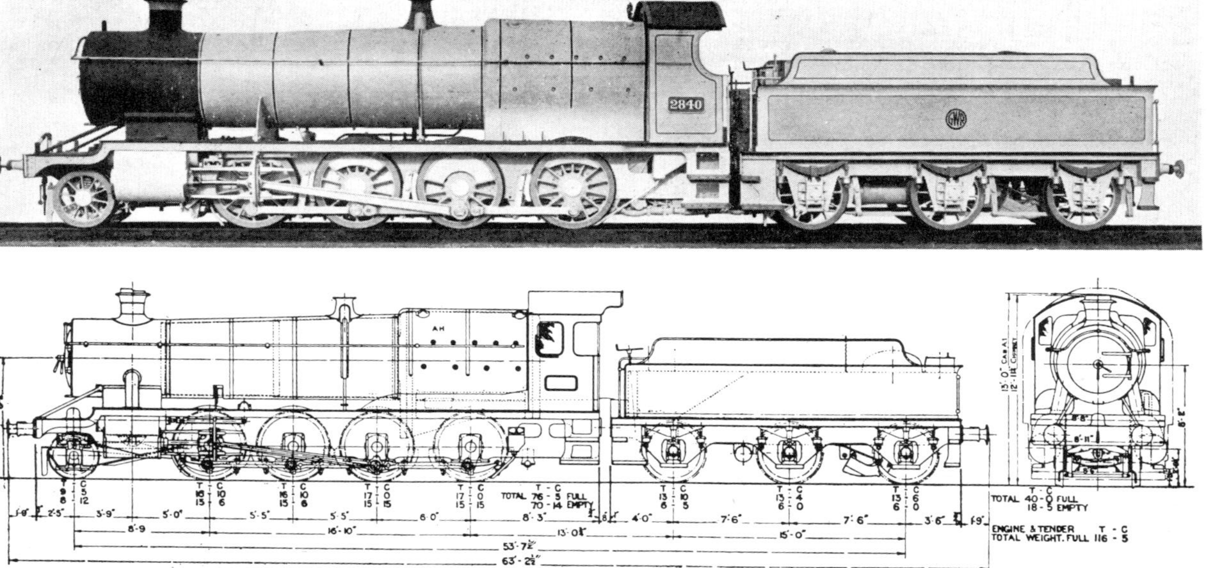

CYLINDERS—Diam., 18½″; Stroke, 30″.
BOILER—Barrel, 14′ 10″; Diam., Outs., 4′ 10 13/16″ and 5′ 6″.
FIREBOX—Outs., 9′ 0″ by 5′ 9″ and 4′ 0″; Ins., 8′ 2 7/16″ by 4′ 9″ and 3′ 2⅝″; Height, 6′ 6⅜″ by 5′ 0⅜″.
TUBES—Superheater Tubes, No. 84: Diam., 1″; Length, 15′ 3⅝″. Fire Tubes, No. 176: Diam., 2″; Length, 15′ 2 7/16″. Fire Tubes, No. 14: Diam., 5⅛″; Length, 15′ 2 7/16″.
TOTAL WEIGHT OF ENGINE—75 tons 10 cwt. Full. 70 tons 2 cwt. Empty.

HEATING SURFACE—Superheater Tubes, 262.62 sq. ft. Fire Tubes, 1,686.6 sq. ft. Firebox, 154.78 sq. ft. Total, 2,104.0 sq. ft.
AREA OF FIREGRATE—27.07 sq. ft.
WHEELS—Pony, 3′ 2″; Coupled, 4′ 7½″.
WATER CAPACITY OF TENDER—3,500 gallons.
WORKING PRESSURE—225 lbs.
TRACTIVE EFFORT—35,380 lbs.
TOTAL WEIGHT OF TENDER—40 tons 0 cwt. Full. 18 tons 5 cwt. Empty.

* Engines 2884—2899 and 3800—3866 are fitted with a new pattern cab. Engine weight—76 tons 5 cwt. Full. 70 tons 14 cwt. Empty.

Serial Numbers :—
4700—4708

4700 Class. Type 2—8—0
(Introduced 1919)

Express Mixed and Perishable Goods Traffic Engines

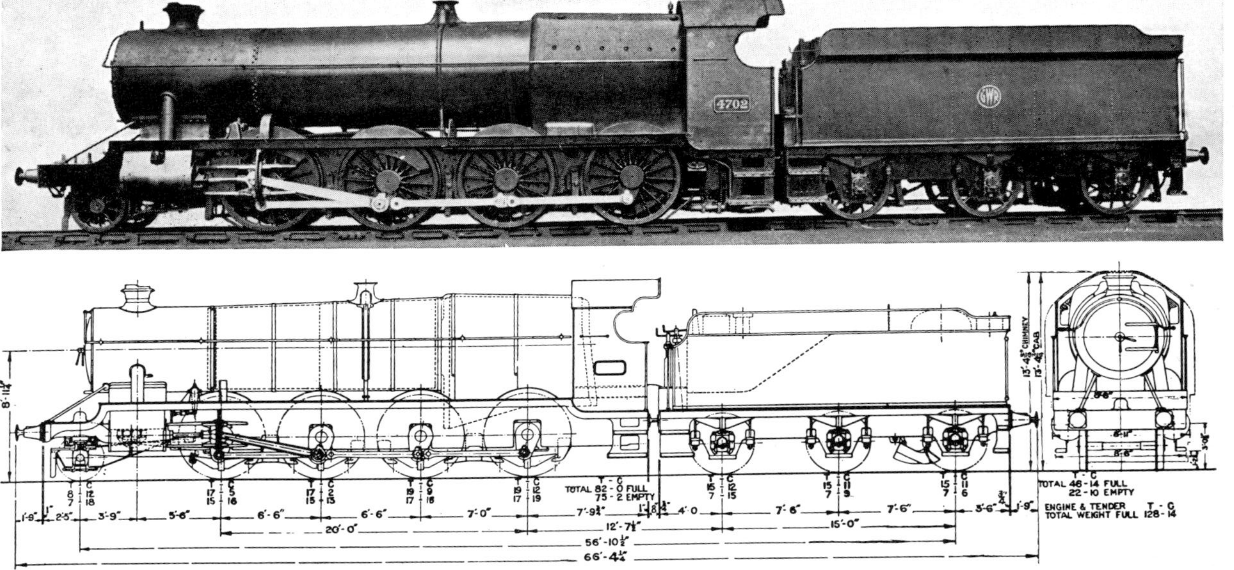

CYLINDERS—Diam., 19″; Stroke, 30″.
BOILER—Barrel, 14′ 10″; Diam. Outs., 5′ 6″ and 6′ 0″
FIREBOX—Outs., 10′ 0″ by 6′ 3″ and 4′ 0″; Ins., 9′ 2⅜″ by 5′ 2⅜″ and 3′ 2⅝″; Height, 6′ 10⅜″ and 5′ 2⅜″.
TUBES—Superheater Tubes, No. 96: Diam., 1″ Length, 15′ 3⅜″; Fire Tubes, No. 218: Diam., 2″; No. 16: Diam., 5⅛″; Length, 15′ 2½″.
TOTAL WEIGHT OF ENGINE—82 tons 0 cwt. Full; 75 tons 2 cwt. Empty.

HEATING SURFACE—Superheater Tubes, 287.53 sq. ft. Fire Tubes, 2,062.35 sq. ft. Firebox, 169.75 sq. ft. Total, 2,519.63 sq. ft.
AREA OF FIREGRATE—30.28 sq. ft. WHEELS—Pony, 3′ 2″; Coupled, 5′ 8″.
WATER CAPACITY OF TENDER—4,000 gallons.
WORKING PRESSURE—225 lbs. TRACTIVE EFFORT—30,460 lbs.
TOTAL WEIGHT OF TENDER—46 tons 14 cwt. Full.
22 „ 10 „ Empty.

Serial Numbers :—
2301—2360 2381—2490
2511—2580 (less gaps)

2301 Class. Type 0—6—0 (Introduced 1883)

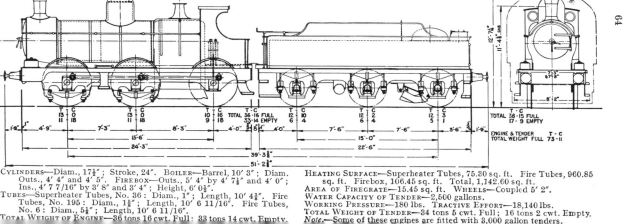

Light Goods Traffic Engines

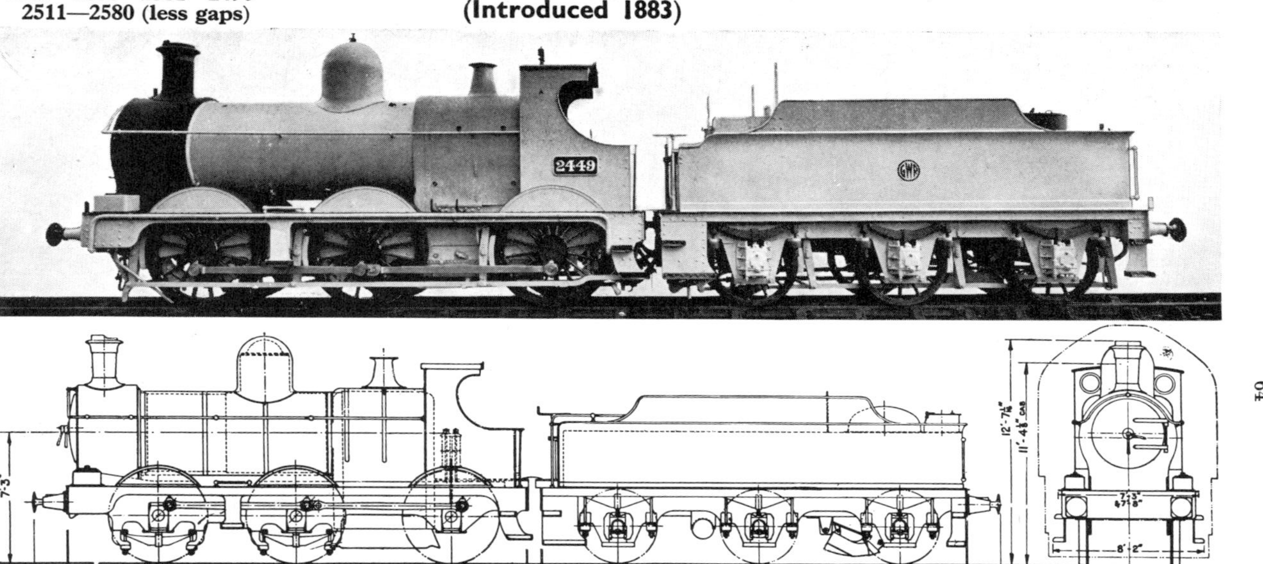

Cylinders—Diam., 17½″; Stroke, 24″. Boiler—Barrel, 10′ 3″; Diam. Outs., 4′ 4″ and 4′ 5″. Firebox—Outs., 5′ 4″ by 4′ 7⅛″ and 4′ 0″; Ins., 4′ 7 7/16″ by 3′ 8″ and 3′ 4″; Height, 6′ 0½″.

Tubes—Superheater Tubes, No. 36: Diam., 1″; Length, 10′ 4⅜″. Fire Tubes, No. 195: Diam., 1⅝″; Length, 10′ 6 11/16″. Fire Tubes, No. 6: Diam., 5⅛″; Length, 10′ 6 11/16″.

Total Weight of Engine—36 tons 16 cwt. Full; 33 tons 14 cwt. Empty.

Heating Surface—Superheater Tubes, 75.30 sq. ft. Fire Tubes, 960.85 sq. ft. Firebox, 106.45 sq. ft. Total, 1,142.60 sq. ft.

Area of Firegrate—15.45 sq. ft. Wheels—Coupled 5′ 2″.

Water Capacity of Tender—2,500 gallons.

Working Pressure—180 lbs. Tractive Effort—18,140 lbs.

Total Weight of Tender—34 tons 5 cwt. Full; 16 tons 2 cwt. Empty.

Note.—Some of these engines are fitted with 3,000 gallon tenders.

Serial Numbers :—
2251—2299

2251 Class. Type 0—6—0
(Introduced 1930)

Light Mixed Traffic Engines

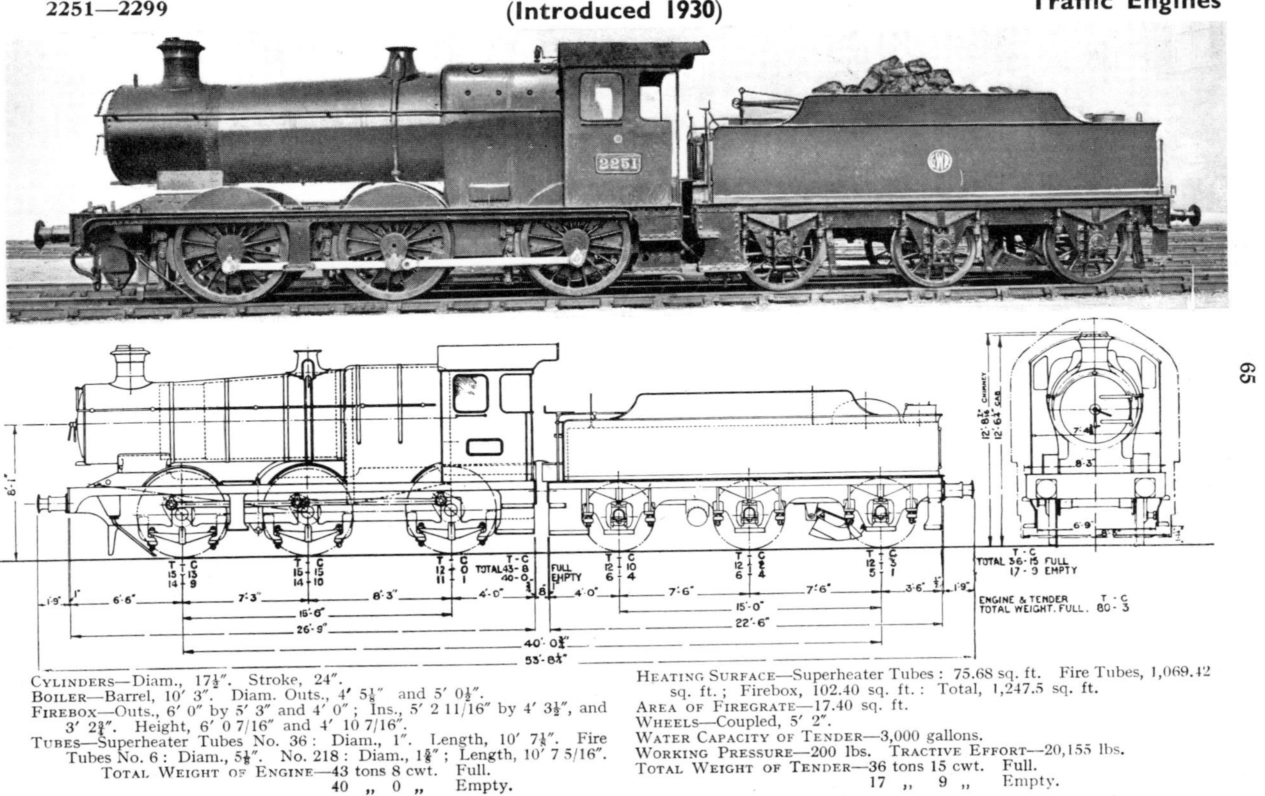

CYLINDERS—Diam., 17½". Stroke, 24".
BOILER—Barrel, 10' 3". Diam. Outs., 4' 5⅛" and 5' 0½".
FIREBOX—Outs., 6' 0" by 5' 3" and 4' 0"; Ins., 5' 2 11/16" by 4' 3½", and 3' 2¾". Height, 6' 0 7/16" and 4' 10 7/16".
TUBES—Superheater Tubes No. 36: Diam., 1". Length, 10' 7⅛". Fire Tubes No. 6: Diam., 5⅛". No. 218: Diam., 1⅝"; Length, 10' 7 5/16".
TOTAL WEIGHT OF ENGINE—43 tons 8 cwt. Full.
40 „ 0 „ Empty.

HEATING SURFACE—Superheater Tubes: 75.68 sq. ft. Fire Tubes, 1,069.42 sq. ft.; Firebox, 102.40 sq. ft.: Total, 1,247.5 sq. ft.
AREA OF FIREGRATE—17.40 sq. ft.
WHEELS—Coupled, 5' 2".
WATER CAPACITY OF TENDER—3,000 gallons.
WORKING PRESSURE—200 lbs. TRACTIVE EFFORT—20,155 lbs.
TOTAL WEIGHT OF TENDER—36 tons 15 cwt. Full.
17 „ 9 „ Empty.

4800 Class. Type $\frac{0-4-2}{T}$

(Introduced 1932)

Serial Numbers :—
4800—4874
5800—5819

These engines are used for working branch passenger services. Nos. 4800 to 4874 are fitted with auto-gear for working with trailer cars.

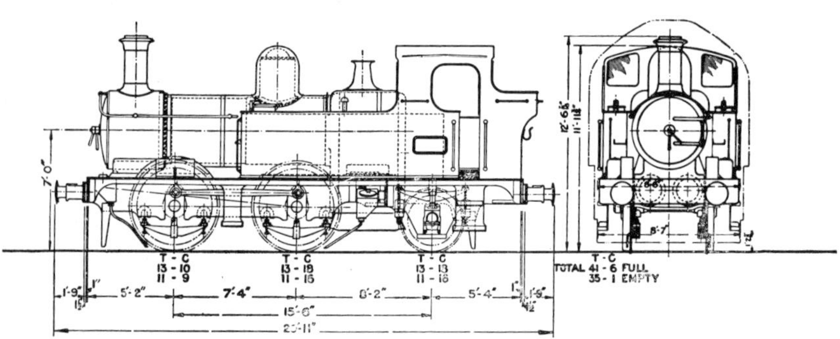

CYLINDERS—Diam. 16″; Stroke 24″.
BOILER—Barrel 10′ 0″; Diam. Outs. 3′ 9⅛″ and 3′ 10″.
FIREBOX—Outs. 4′ 6″ by 4′ 0″; Ins. 3′ 9 15/16″ by 3′ 4″. Height 5′ 5½″.
TUBES—No. 2: Diam. 5⅛″. No. 193: Diam. 1⅝″. Length 10′ 3 3/16″.

HEATING SURFACE—Tubes 869.8 sq. ft. Firebox 83.2 sq. ft. Total, 953.0 sq. ft.
AREA OF FIREGRATE—12.8 sq. ft.
WHEELS—Coupled, 5′ 2″: Trailing, 3′ 8″.
WATER CAPACITY OF TANKS—800 gallons.
WORKING PRESSURE—165 lbs. sq. in.
TRACTIVE EFFORT—13 900 lbs.

TOTAL WEIGHT OF ENGINE—41 tons 6 cwt. Full.
35 ,, 1 ,, Empty.

5400 Class. 0—6—0 / T

(Introduced 1931)

Serial Numbers :—
5400—5424
***6400—6439**
***7400—7429**

These engines are of heavier type than the 4800 class and are fitted with auto-gear for working with trailer cars on branch passenger services, except Nos. 7400—7429.

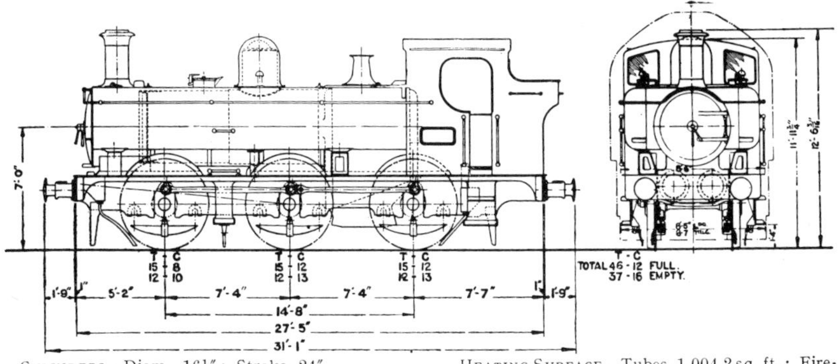

CYLINDERS—Diam., $16\frac{1}{2}$"; Stroke, 24".
BOILER—Barrel, 10' 6"; Diam. Outs., 4' $2\frac{1}{8}$" and 4' 3".
FIREBOX—Outs., 5' 6" by 4' 0"; Ins., 4' 9 15/16" by 3' 4"; Height, 5' $4\frac{3}{4}$" and 3' $9\frac{1}{4}$".
TUBES—No. 2: Diam., $5\frac{1}{8}$"; No. 213: Diam., $1\frac{5}{8}$"; Length, 10' 9 3/16".

HEATING SURFACE—Tubes, 1,004.2 sq. ft.: Firebox, 81.8 sq. ft. Total, 1,086 sq. ft.
AREA OF FIREGRATE—16.76 sq. ft.
WHEELS—Coupled, 5' 2".
WATER CAPACITY OF TANKS—1,100 gallons.
WORKING PRESSURE—165 lbs. sq. in.
TRACTIVE EFFORT—14,780 lbs.

TOTAL WEIGHT OF ENGINE—46 tons 12 cwt. Full; 37 tons 16 cwt. Empty.

*Fitted with 4' $7\frac{1}{2}$" wheels, and have a working pressure of 180 (7400 class) and 165 (6400 class) lbs. sq. in; and tractive effort of 18,010 lbs. (7400 class) and 16,510 lbs. (6400 class).

5700 Class. Type $\frac{0-6-0}{T}$

(Introduced 1929)

Serial Numbers :—

5700—5799	7700—7799
6700—6749	8700—8799
9711—9799	3700—3799
3600—3699	4600—4699
9600—9650	

Light goods and shunting engines.

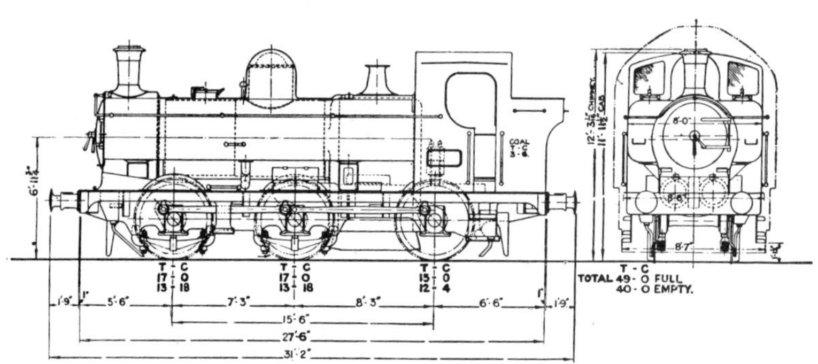

Cylinders—Diam., $17\frac{1}{2}''$; Stroke, 24″.
Boiler—Barrel, 10′ 3″; Diam. Outs., 4′ $3\frac{7}{8}''$ and 4′ 5″.
Firebox—Outs., 5′ 4″ by 4′ $7\frac{3}{4}''$ and 4′ 0″: Ins., 4′ 7 3/16″ by 3′ 8″ and 3′ $3\frac{3}{4}''$; Height, 6′ $0\frac{1}{8}''$.
Tubes—No. 2: Diam., $5\frac{1}{8}''$. No. 233: Diam., $1\frac{5}{8}''$ Length, 10′ 6 13/16″.

Heating Surface—Tubes, 1,075.7 sq. ft. Firebox, 102.3 sq. ft. Total, 1,178.0 sq. ft.
Area of Firegrate—15.3 sq. ft.
Wheels—Coupled, 4′ $7\frac{1}{2}''$.
Water Capacity of Tanks—1,200 gallons.
Working Pressure—200 lbs.
Tractive Effort—22,515 lbs.

Total Weight of Engine—49 tons Full.
40 ,, Empty.

1366 Class. Type 0—6—0 T

(Introduced 1934)

Serial Numbers :—
1366—1371

Dock shunting engines.

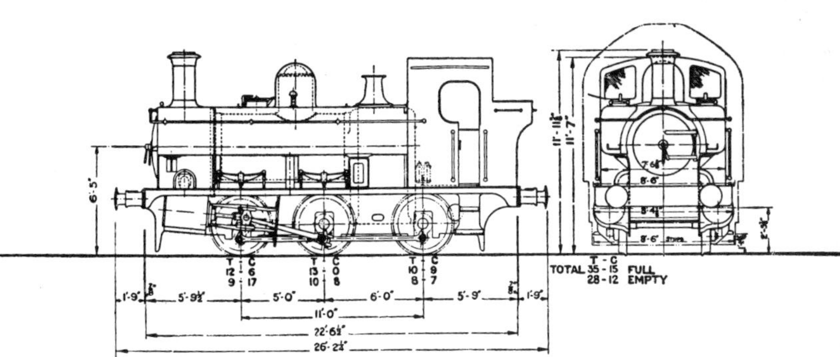

Cylinders—Diam., 16″ ; Stroke, 20″.
Boiler—Barrel, 8′ 2″ ; Diam. Outs., 3′ 10″.
Firebox—Outs., 3′11″ by 4′ 0″ ; Ins., 3′ 2 15/16″ by 3′ 4″ : Height, 5′ 3½″.
Tubes—No. 2. Diam., 5⅛″. No. 193 : Diam. 1⅝″. Length, 8′ 5 3/16″.
Heating Surface—Tubes, 715 sq. t. Firebox, 73 sq. ft. Total, 788 sq. ft.

Area of Firegrate—10.7 sq. ft.
Wheels—Leading, 3′ 8″ ; Driving, 3′ 8″ ; Trailing, 3′ 8″.
Water Capacity of Tank—830 gallons.
Working Pressure—165 lbs.
Tractive Effort—16,320 lbs.

Total Weight of Engine—35 tons 15 cwt. Full.
28 ,, 12 ,, Empty.

3150 Class. Type 2—6—2 T

(Introduced 1907)

Serial Numbers :—
3150—3190
3100—3104*

Heavy surburban passenger traffic engines.

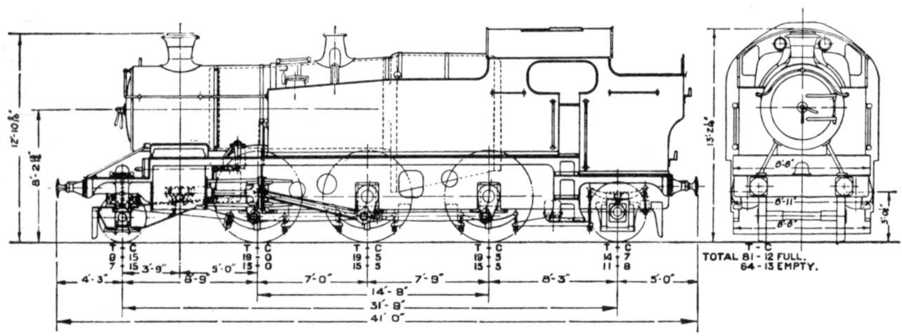

Cylinders—Diam., 18½"; Stroke 30".
Boiler—Barrel, 11' 0"; Diam. Outs., 4' 10¾" and 5' 6".
Firebox—Outs., 7' 0" by 5' 9" and 4' 0"; Ins., 6' 2½" by 4' 8⅞" and 3' 2⅝"; Height, 6' 6⅜" and 5' 0⅜".
Tubes—Superheater Tubes, No. 84: Diam., 1"; Length, 11' 5⅜". Fire Tubes, No. 235: Diam., 1⅝": Length, 11' 4 7/16". Fire Tubes, No. 14: Diam., 5⅛"; Length, 11' 4 7/16".

Heating Surface—Superheater Tubes, 191.88 sq. ft. Fire Tubes, 1,349.64 sq. ft. Firebox, 128.72 sq. ft. Total, 1,670.24 sq. ft.
Area of Firegrate—20.56 sq. ft.
Wheels—Pony Truck, 3' 2"; Coupled, 5' 8"; Radial Truck, 3' 8".
Water Capacity of Tanks—2,000 gallons.
Working Pressure—200 lbs.: 225 lbs. ("3100" class).
Tractive Effort—25,670 lbs.: 31,170 lbs. ("3100" class).

Total Weight of Engine—81 tons 12 cwt. Full.
64 ,, 13 ,, Empty.

*3100—3104 have 5' 3" coupled wheels.
Total Weight of Engine—81 tons 9 cwt. Full.
65 ,, 15 ,, Empty.

5100 Class. Type 2—6—2 T

(Introduced 1929)

Serial Numbers :—
5100—5199
6100—6169
4100—4139
8100—8109*

A development of the 3100 class suburban passenger traffic engines.

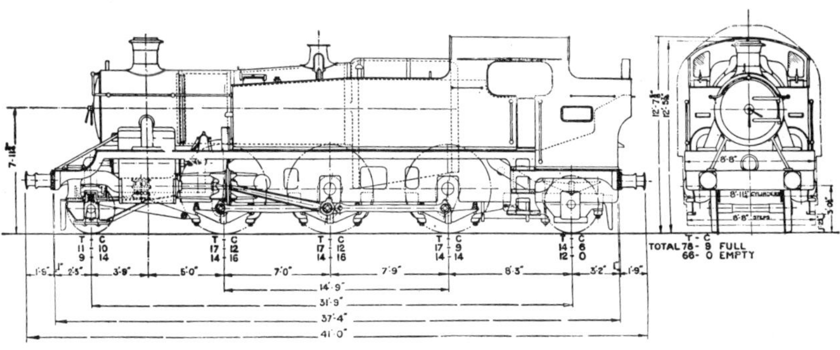

CYLINDERS—Diam., 18″ : Stroke, 30″.
BOILER—Barrel, 11′ 0″. Diam. Outs., 4′ $5\frac{1}{8}$″ and 5′ $0\frac{1}{2}$″.
FIREBOX—Outs., 7′ 0″ by 5′ 3″ and 4′ 0″ ; Ins., 6′ 2 11/16″ by 4′ $3\frac{1}{2}$″ and 3′ $2\frac{3}{4}$″. Height, 6′ 0 7/16″ and 5′ 0 7/16″.
TUBES—Superheater Tubes No. 36 ; Diam., 1″ · Length, 11′ $5\frac{3}{8}$″. Fire Tubes No. 6 : Diam., $5\frac{1}{8}$″. No. 218: Diam., $1\frac{5}{8}$″: Length, 11′ 4 5/16″.

TOTAL WEIGHT OF ENGINE—
78 tons 9 cwt. Full.
66 ,, 0 ,, Empty.

HEATING SURFACE—Superheater Tubes, 82.30 sq. ft. Fire Tubes, 1,145. sq. ft. Firebox, 121.80 sq. ft. Total, 1,349.10 sq. ft.
AREA OF FIREGRATE—20.35 sq. ft.
WHEELS—Pony Truck, 3′ 2″ ; Coupled, 5′ 8″ ; Radial Truck, 3′ 8″.
WATER CAPACITY OF TANKS—2,000 gallons.
WORKING PRESSURE—200 lbs. : 225 lbs. (6100 and 8100 classes).
TRACTIVE EFFORT—24,300 lbs. (5100 class).
27,340 ,, (6100 ,,).
28,165 ,, (8100 ,,).

*Engines 8100—8109 have 5′ 6″ coupled wheels.
TOTAL WEIGHT OF ENGINE—76 tons 11 cwt. Full.
65 ,, 13 ,, Empty.

4200 Class. Type 2—8—0 T

(Introduced 1910)

Serial Numbers :—
4200—4299
5200—5264

Heavy coal traffic engines.

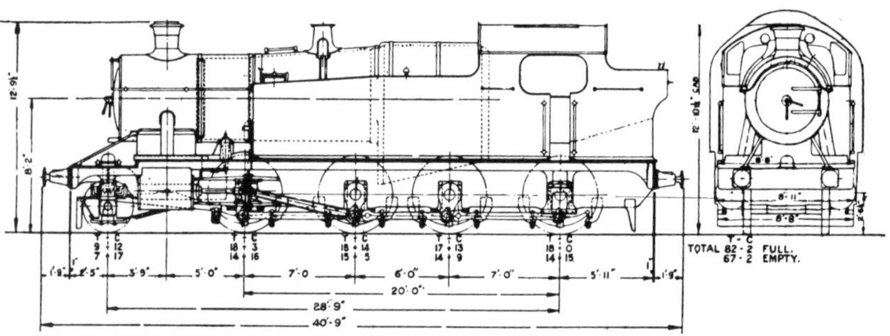

Cylinders—Diam., 19″; Stroke, 30″.
Boiler—Barrel, 11′ 0″; Diam., Outs., 4′ 10¾″ and 5′ 6″.
Firebox—Outs., 7′ 0″ by 5′ 9″ and 4′ 0″; Ins., 6′ 2½″ by 4′ 8⅞″ and 3′ 2⅝″; Height, 6′ 6⅜″ and 5′ 0⅜″.
Tubes—Superheater Tubes, No. 84: Diam., 1″; Length, 11′ 5⅜″. Fire Tubes, No. 235: Diam., 1⅝″; Length, 11′ 4 7/16″. Fire Tubes, No. 14: Diam., 5⅛″; Length 11′ 4 7/16″.

Heating Surface—Superheater Tubes, 191.88 sq. ft. Fire Tubes, 1,349.64 sq. ft. Firebox, 128.72 sq. ft. Total, 1,670.24 sq. ft.
Area of Firegrate—20.56 sq. ft.
Wheels—Pony Truck, 3′ 2″; Coupled, 4′ 7½″.
Water Capacity of Tank—1,800 gallons.
Working Pressure—200 lbs.
Tractive Effort—33,170 lbs.

Total Weight of Engine—82 tons 2 cwt. Full.
67 ,, 2 ,, Empty.

7200 Class. Type $\frac{2\text{—}8\text{—}2}{\text{T}}$

(Introduced 1934)

Serial Numbers :— 7200—7253

Conversions from 2—8—0 T type for heavy main line coal traffic, to replace 2600 class.

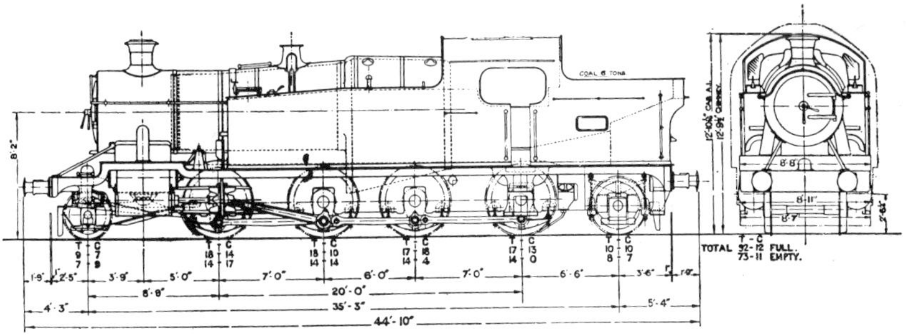

Cylinders—Diam., 19″; Stroke, 30″.
Boiler—Barrel, 11′ 0″; Diam. Outs., 4′ 10¾″ and 5′ 6″.
Firebox—Outs., 7′ 0″ by 5′ 9″ and 4′ 0″; Ins., 6′ 2½″ by 4′ 8⅞″ and 3′ 2⅝″; Height, 6′ 6⅜″ and 5′ 0⅜″.
Tubes—Superheater Tubes, No. 84: Diam., 1″; Length, 11′ 5⅜″. Fire Tubes, No. 14: Diam., 5⅛″; No. 235: Diam., 1⅝″: Length, 11′ 4 7/16″.

Heating Surface—Superheater Tubes, 191.88 sq. ft. Fire Tubes. 1.349.64 sq. ft.; Firebox, 128.72 sq. ft.; Total, 1,670.24 sq. ft.
Area of Firegrate—20.56 sq. ft.
Wheels—Pony Truck, 3′ 2″; Coupled, 4′ 7½″; Radial Truck, 3′ 8″.
Water Capacity of Tanks—2,500 gallons.
Working Pressure—200 lbs. sq. in.
Tractive Effort—33,170 lbs.

Total Weight of Engine—92 tons 12 cwt. Full.
73 ,, 11 ,, Empty

4500 Class. Type 2—6—2 T

(Introduced 1906)

Serial Numbers :—
4500—4599
5500—5574

Light branch line passenger traffic engines.

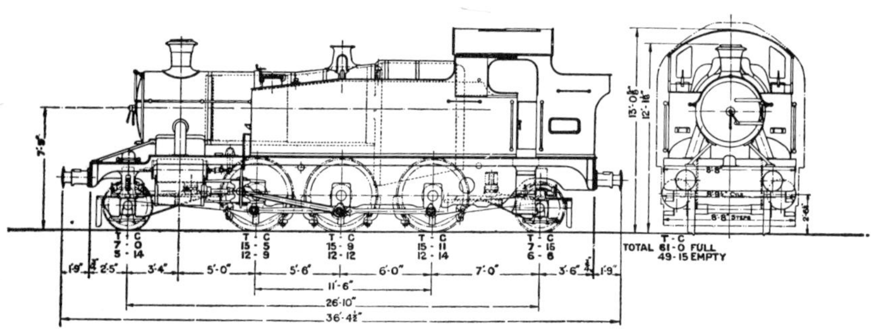

CYLINDERS—Diam., 17″; Stroke, 24″.

BOILER—Barrel, 10′ 6″; Diam. Outs., 4′ 2″ and 4′ 9½″.

FIREBOX—Outs., 5′ 10″ by 5′ 0″ and 4′ 0″; Ins., 5′ 0 11/16″ by 4′ 0″ and 3′ 2¾″; Height, 5′ 8″ and 4′ 8″.

TUBES—Superheater Tubes, No. 36: Diam., 1″; Length, 10′ 10⅛″. Fire Tubes, No. 196: Diam., 1⅝″; Length, 10′ 10 5/16″. Fire Tubes, No. 6: Diam. 5⅛″; Length, 10′ 10 5/16″.

HEATING SURFACE—Superheater Tubes, 73.80 sq. ft. Fire Tubes, 992.51 sq. ft. Firebox, 94.25 sq. ft. Total, 1,160.56 sq. ft.

AREA OF FIREGRATE—16.6 sq. ft.

WHEELS—Pony Truck, 3′ 2″; Coupled, 4′ 7½″ Pony Truck, 3′ 2″.

WATER CAPACITY OF TANK—1,300 gallons.

WORKING PRESSURE—200 lbs.

TRACTIVE EFFORT—21,250 lbs.

TOTAL WEIGHT OF ENGINE—61 tons 0 cwt. Full.
49 ,, 15 ,, Empty.

5600 Class. Type $\frac{0-6-2}{T}$

(Introduced 1924)

Serial Numbers :—
5600—5699
6600—6699

Heavy mixed traffic engines for work in South Wales coalfields.

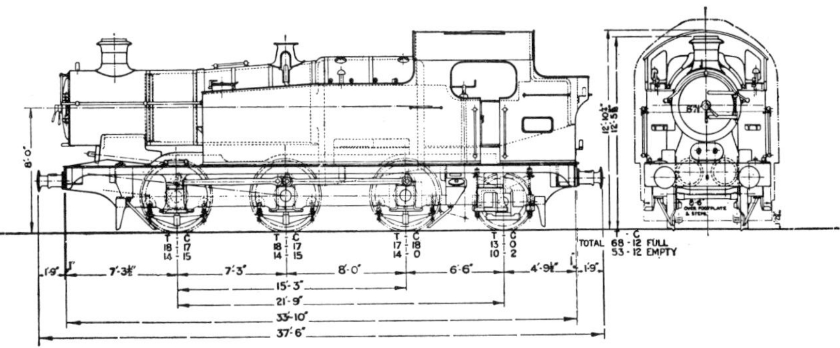

CYLINDERS—Diam., 18″ ; Stroke, 26″.

BOILER—Barrel, 11′ 0″ ; Diam. Outs., 4′ 5⅛″ and 5′ 0½″.

FIREBOX—Outs., 7′ 0″ by 5′ 3″ and 4′ 0″ ; Ins., 6′ 2 11/16″ by 4′ 3½″ and 3′ 2¾″ ; Height, 6′ 0 7/16″ and 5′ 0 7/16″.

TUBES—Superheater Tubes, No. 36 : Diam., 1″ ; Length, 11′ 5⅜″ ; Fire Tubes, No. 6 : Diam. 5⅛″ No. 218 : Diam. 1⅝″, Length, 11′ 4 5/16″.

HEATING SURFACE—Superheater Tubes, 82.30 sq. ft. Fire Tubes, 1,145 sq. ft. Firebox, 121.80 sq. ft. Total, 1,349.10 sq. ft.

AREA OF FIREGRATE—20.35 sq. ft.

WHEELS—Coupled, 4′ 7½″ ; Radial, 3′ 8″.

WATER CAPACITY OF TANKS—1,900 gallons.

WORKING PRESSURE—200 lbs.

TRACTIVE EFFORT—25,800 lbs.

TOTAL WEIGHT OF ENGINE—68 tons 12 cwt. Full.
53 ,, 12 ,, Empty.

1901 Class. Type $\frac{0\text{—}6\text{—}0}{\text{T}}$

(Introduced 1874)

Serial Numbers :—	
850—873	987—998
1216—1227	1901—2020

For light branch and shunting work.

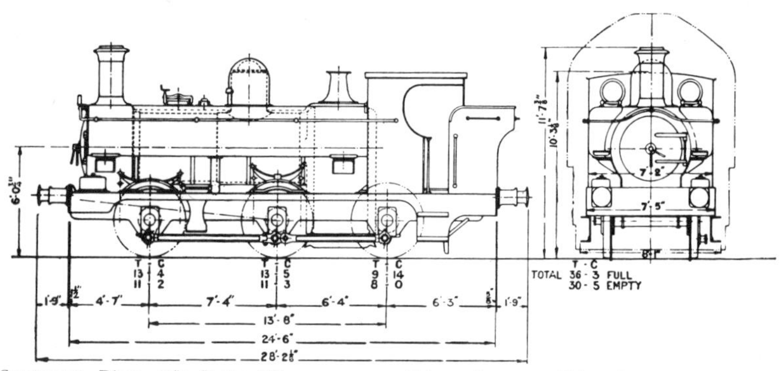

CYLINDERS—Diam., 16″; Stroke, 24″.
BOILER—Barrel, 10′ 0″; Diam., Outs., 3′ $9\frac{1}{8}$″ and 3′ 10″.
FIREBOX—Outs., 4′ 0″ by 4′ 0″; Ins., 3′ 3 15/16″ by 3′ 4″. Height, 5′ $5\frac{1}{2}$″.
TUBES—No. 192: Diam., $5\frac{1}{8}$″. No. 193: Diam. $1\frac{5}{8}$″. Length, 10′ 3 3/16″.

HEATING SURFACE—Tubes, 870.86 sq. ft. Firebox, 76.28 sq. ft. Total, 947.14 sq. ft.
AREA OF FIREGRATE—11.16 sq. ft.
WHEELS—Coupled, 4′ $1\frac{1}{2}$″.
WATER CAPACITY OF TANKS—800 gallons.
WORKING PRESSURE—165 lbs.
TRACTIVE EFFORT—17,410 lbs.

Locomotive No. 6003—" King George IV "

Diesel Rail Cars

Number 19

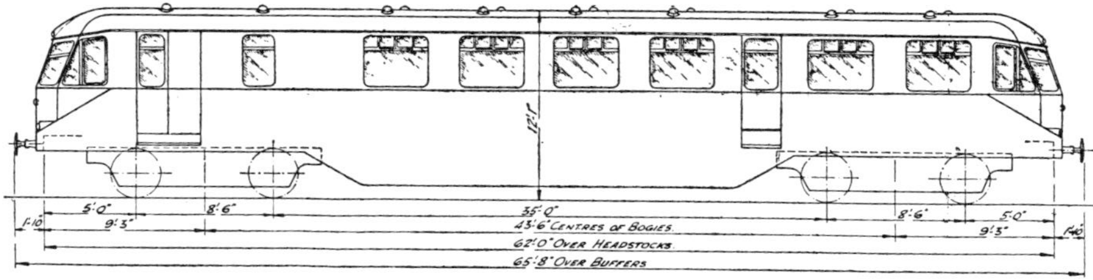

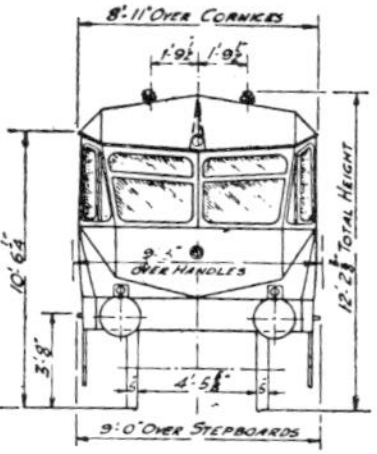

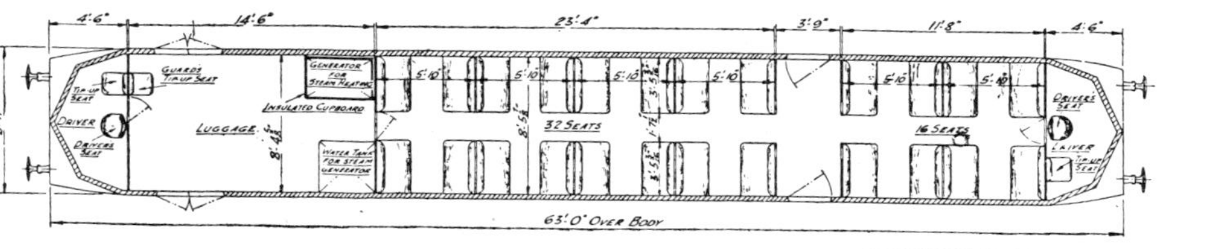

Diesel Rail Cars
(Introduced 1934)

Serial Numbers :—
1—38

Passenger and Parcels traffic

No. 1.

This car was put into service in February, 1934. It is equipped with a single engine of 121 B.H.P. with a four-speed and reverse gearbox. It has seating accommodation for sixty-nine passengers. There is a small driver's and luggage compartment at one end and a driver's compartment at the other. The body of this car was built by Messrs. Park Royal Coachworks Ltd.

Nos. 2 to 4.

These cars commenced service in July, 1934 and are fitted with two 121 B.H.P. engines, having four-speed and reverse gearboxes. The brake is of the Lockheed Vacuum-hydraulic type.

They are equipped with a small buffet compartment at one end with four seats, and between this and the main compartment are two lavatory compartments. The seating capacity of the main compartments is forty, giving a total of forty-four seats. There is a driver's compartment at each end.

Like Car No. 1, the bodies were made by Messrs. Park Royal Coachworks Ltd.

Nos. 5 to 17.

The first of these cars went into service in July, 1935. They are fitted with two engines with a total of 242 B.H.P., and gearboxes similar to Nos. 2 to 4. A vacuum brake is fitted with one cylinder operating on each wheel through brake-drums.

The bodies were constructed by Messrs. The Gloucester Carriage and Wagon Co., Ltd. The seating capacity of cars Nos. 5 to 9 and 13 to 16 is seventy and for cars Nos. 10 to 12, which have a lavatory compartment, sixty-three. All these cars are provided with a driver's compartment at each end and (with the exception of No. 17 which is fitted for carrying parcels and has no passenger accommodation) are provided with a small luggage and guard's compartment.

No. 18.

This car was designed to draw a tail load of sixty tons and has standard railway type buffers and drawgear. It is equipped with two engines with a total B.H.P. of 242. The vacuum-brake operates through brake-drums. The body was built by Messrs. The Gloucester Carriage and Wagon Co. Ltd. It was put into service in January, 1937.

Nos. 19 to 33.

The first of these cars was put into service in June, 1940. They have two engines, each of 105 B.H.P., and are fitted with pre-selector gears. They are designed to take a tail load of sixty tons and are fitted with standard railway buffers and drawgear, and with clasp-type brakes and brake blocks operated by two vacuum cylinders per car. The seating capacity is forty-eight and there is a driver's compartment at each end and a luggage and brake compartment. The bodies were constructed at Swindon Works.

No. 34.

No. 34 is similar to Nos. 19 to 33 except that it is arranged for parcels traffic and has no passenger accommodation.

Nos. 35 to 38.

These form two twin-coach sets, one coach having a seating capacity of sixty and the other forty-four, giving a total of 104 passengers. One coach in each set has, besides the passenger compartments, a luggage and guard's compartment and a lavatory compartment. The other coach is fitted with a buffet with counter for snack meals and a luggage and guard's compartment. The bodies differ from Nos. 19 to 33 in the fact that there is only one driver's compartment on each vehicle. these being situated at the extreme ends of the set. Each car is provided with engines and transmission gear similar to the single cars Nos. 19 to 34.

The engines and chassis of all these cars were built by Messrs. The Associated Equipment Company at their Southall Works.

"A" Erecting Shop, G.W.R. Locomotive Works, Swindon.

SWINDON

Where G.W.R. Locomotives are made

THE Swindon Works of the Great Western Railway comprise one of the largest railway establishments for the construction and repair of locomotives, carriages and wagons in the world, and since the acquisition of the earliest engines for opening the railway in 1838 (with but few exceptions) all the locomotives of the Great Western Railway have been constructed there.

The Works are situated on both sides of the main line to Bristol and the West of England, and give employment to about 12,000 men. The Locomotive Works include Fitting, Erecting, Boiler-making, and Machine Shops ; Iron and Brass Foundries ; Smiths', Stamping, Tinsmiths' and Coppersmiths' Shops.

On the average two new locomotives are constructed weekly, whilst the capacity for overhaul and repairs is approximately 1,000 locomotives per annum.

Steady expansion of the Works has been almost continuous, particular attention being paid to the equipment which, at the present time, comprises the most up-to-date tools and appliances that can be found in the world, including a number of most fascinating automatic machines which turn out complete units used in locomotive construction with the minimum of human supervision.

This equipment proved invaluable during the War years when much of the machinery was kept running in the production of munitions of various kinds for seven days a week for long periods. War equipment was produced for all three of the fighting services and included, among a multitude of items, landing craft and midget submarines for the Royal Navy, shells of various calibres, guns and armoured vehicles for the Army, and radiolocation equipment and bombs for the Royal Air Force.

When the full story of the Great Western Railway's contribution to victory is told, the achievements of Swindon Works will be no mean portion of a glorious record of high endeavour.

G.W.R. ENGINES AT WAR

DESPITE the unprecedented demand for locomotive power during the War period 1939–45 for the conveyance of personnel of the fighting forces, munitions and all the *impedimenta* of war, the Great Western Railway handed over to the Government, 108 goods engines of the " 2301 " class and, as far as is known, 79 of them were sent for duty on the Continent prior to the Dunkirk evacuation.

These engines had their G.W.R. number plates removed and W.D. numbers substituted on the sides of the cabs, and the letters W↑D on the tenders. They were painted black, and most of them were fitted with Westinghouse Brake gear.

Ten of the engines (W.D. 177 to 180 and W.D. 195 to 200) were equipped with condensing gear and pannier tanks, carried on each side of the boilers, into which exhaust steam could be directed at will from the blast pipe and discharged beneath the ashpan. The two reservoirs of the condensing engines were placed vertically at the left-hand side of the smoke-box, the Westinghouse pump being on the right-hand side. Water-lifting apparatus (Whites') was installed in the cabs to which suction hose was connected (normally kept at the back of the tender) to enable water to be picked up from any convenient pond or stream to replenish tenders or condensing tanks.

There is at present no information available as to the fate of these engines, and it is not known whether any of them were casualties by enemy action, although it is believed that some of them found their way to the Middle East after the collapse of France. It is interesting to note that a number of these engines also served in France and Salonika in the 1914–18 War.

Between July, 1940 and March, 1944, no fewer than 69 G.W.R. engines were the victims of the *Luftwaffe* in this country and in 23 cases the damage sustained was graded as *heavy*. Two engines, No. 4911, *Bowden Hall* (4–6–0), and No. 1729 (0–6–0T), suffered so badly at Keyham, Plymouth on April 29th, 1941, and at Castle Cary on March 30th, 1942 respectively, that they had to be condemned and broken up, as also had the tender of No. 4358 (2–6–0), damaged at Bristol on December 6th, 1940.

In raids on August 20th, 1940, sixteen G.W.R. engines were damaged by enemy air attack, some heavily ; no fewer than fifteen of them at Newton Abbot. Twelve locomotives were similarly damaged at Bristol between August, 1940 and March, 1941.

Besides bomb damage, a number of engines suffered from enemy cannon and machine-gun fire, and two had the misfortune to run into large bomb craters. One engine, No. 6433 (0–6–0T), earned two " wound stripes," being heavily damaged at Cardiff in August, 1940, and (after being repaired and put back into service) by further air attack at Cardiff in January of the following year.

Bristol to Paddington Express. Locomotive No. 6000 "King George V"

TRAIN SPEEDS

THE Great Western Railway had a world-wide reputation for the speed of its trains as early as the middle of the last century, and since that time the initials *G.W.R.* have been regarded as a symbol of speedy transit.

During the years of war it was necessary in the national interest to impose restrictions of speed on trains for a variety of reasons and, although hostilities have ceased, post-war conditions do not yet permit of a return to 1939 standards. It must be remembered that many hundreds of skilled railwaymen still await demobilization, inferior coal has to be used on account of the national shortage, and large numbers of locomotives and other rolling stock, overworked during war years, await overhaul and repairs.

It may be of interest, however, to recall some of the train speeds prior to the outbreak of war in September, 1939. The regular train service then embraced a large number of trains which ran for considerable distances at high average speeds, including the record run of "Cheltenham Flyer" (2.40 p.m. from Cheltenham) over the 77¼ miles from Swindon to Paddington in 65 minutes, or an average speed of 71.4 miles an hour. Whilst that particular run was unique, here are some other fast runs booked at speeds of over a mile-a-minute, start to stop, for a distance exceeding 50 miles.

		Distance Miles	Time Mins.	Speed m.p.h.
10. 0 a.m.	Paddington to Bristol	118¼	105	67·6
4.30 p.m.	Bristol to Paddington	117½	105	67·1
10.10 a.m. 5.35 p.m.	Oxford to Paddington	63½	60	63·5
11.15 a.m. 1.15 p.m.	Paddington to Bath	106¾	102	62·8
10.18 a.m.	Westbury to Paddington	95½	92	62·3
5. 5 p.m.	Paddington to Bath	106¾	103	62·1
9. 2 a.m.	Kemble to Paddington	91	88	62·0
12. 0 noon.	Paddington to Exeter	173¾	170	61·3
1.40 a.m.	Paddington to Taunton	143	140	61·3
10,30 a.m.W	Paddington to Taunton (Slip)	143	140	61·3
10.30 a.m.W	Paddington to Exeter	173¾	171	60·9
9.44 a.m.	High Wycombe to Leamington (Slip)	60¾	60	60·7
5. 0 p.m.	Paddington to Kemble	91	90	60·6
8.27 a.m.	Chippenham to Paddington	94	93	60·6
3.17 p.m.	Oxford to Paddington	63½	63	60·4
3.30 p.m.	Paddington to Westbury	95½	95	60·3
1.13 p.m. 4.18 p.m.	Swindon to Paddington	77¼	77	60·2
9.24 a.m.	Didcot to Paddington	53¼	53	60·1

W = Winter Time Table.

Locomotives ancient and modern—" North Star " and " King George V "

THE GROWTH OF LOCOMOTIVE POWER

THE development of the railway locomotive is a story of ever-increasing tractive power. Whilst comparatively high speeds were attained on the Great Western Railway as far back as the 'forties, the weights of the trains hauled in those days were but a fraction of those with which the modern locomotives have to deal. From a few small carriages the train load has grown to fifteen or more 60-ft. coaches, including dining and kitchen cars, and having an aggregate weight of 500 tons behind the tender. Such trains are to-day hauled at high average speed on long " non-stop " runs.

North Star (2–2–2), one of the first locomotives supplied to the Company—built by Messrs. Robert Stephenson & Co. in 1837, and originally intended for the New Orleans Railway of America—was the engine which gave the most satisfactory performance at the public opening of the line in June, 1838. She was considerably altered in construction at Swindon Works in 1854 and remained in active service until 1870.

Another old broad-gauger of fame was *Lord of the Isles* (4–2–2), built, like practically all G.W.R. locomotives (after the very earliest days), at the Company's locomotive works at Swindon. Produced in 1851, she was probably the fastest and most powerful broad-gauge locomotive that ever existed. She ran 789,300 miles with her original boiler intact. Wonderful stories are told of speeds obtained by these old eight-foot single-wheelers. Trains hauled by them were actually booked to leave Didcot (53 miles) fifty-seven minutes after departure from Paddington.

The present *North Star*, built in 1906, as 4–4–2 type, was converted to 4–6–0 in 1909. In this form she was typical of a famous class of four-cylinder engines which may be said to have marked an epoch in railway locomotive construction.* Large numbers of these engines are included in the G.W.R. " fleet," and they have rendered admirable

* *North Star* was in 1931 reconstructed as a " Castle " class locomotive.

service. The "Stars" and other types closely resembling them were responsible for creating and regularly maintaining such speedy "non-stop" runs as that from Paddington to Plymouth (225¾ miles), a world's record for over twenty-one years.

Caerphilly Castle made her *début* in August, 1923, and was followed by some thirty sister locomotives of the "Castle" class. These engines, which were a development of the "Star" class, embodied many new and distinct features, including a larger boiler, and had a weight (with tender) in working order of about 126 tons. When exhibited at the British Empire Exhibition at Wembley in 1924, *Caerphilly Castle,* then the most powerful passenger locomotive in the Kingdom, attracted a continuous stream of admirers. A sister engine, *Windsor Castle,* was driven by H.M. King George V when he visited Swindon Works.

In the year 1927 the first of the now world-famous "King" class locomotives appeared. Thirty of these super-locomotives have been constructed at Swindon Works. The "Kings" are also of the four-cylinder 4–6–0 type, and, at 85 per cent. boiler pressure, have a tractive effort of 40,300 lbs., which makes them amongst the most powerful passenger locomotives in Great Britain to-day.

These engines are named after the Kings of England, and it was fitting that the first of the class completed should bear the name of the then reigning monarch, King George V. Considerable interest is taken in British locomotive practice in America, and there was nothing but admiration for *King George V* by visitors who flocked to the Baltimore and Ohio Centenary Exhibition in 1927, where that engine was on view with the old *North Star.*

The question may be asked: "What are the advantages of these high-powered locomotives?" An example of what has been accomplished by their use is furnished by the journey from Paddington to Exeter, a distance of 173¾ miles, which in normal times, is covered by the Cornish Riviera and Torbay Expresses in 170 minutes—an average speed of 61.3 miles per hour, compared with the previous fastest time of 180 minutes or an average speed of 57.9 miles per hour.

It should, perhaps, be added that no engines of greater power than those of the " King " class have been produced by the Great Western Railway since 1928 and these locomotives have proved adequate for all requirements and to meet those which may be made upon the Locomotive Department in the immediate post-war time-table.

THE GROWTH OF LOCOMOTIVE POWER

GREAT WESTERN RAILWAY PASSENGER LOCOMOTIVE DEVELOPMENT

	1837 North Star	1851 Lord of the Isles	1909 North Star	1923 Caerphilly Castle	**1927 King George V**
CYLINDERS	(2)	(2)	(4)	(4)	**(4)**
Diam. and Stroke ..	16×16	18×24	$14\frac{1}{4}$×26	16×26	**$16\frac{1}{4}$×28**
DRIVING WHEEL: Diameter	7ft. 0in.	8ft. 0in.	6ft. $8\frac{1}{2}$in.	6ft. $8\frac{1}{2}$in.	**6ft. 6in.**
BOILER PRESSURE Lbs. per sq. in. ..	50	140	225	225	**250**
HEATING SURFACE: Total—Sat. and Sup. sq. ft. ..	694·7	1,750	2,014·4	2,312	**2,514**
GRATE AREA: Sq. ft. ..	13·62	24·0	27·07	30·28	**34·3**
TRACTIVE EFFORT: m.e.p.—85% B.P.	2,070	9,640	25,085	31,625	**40,300**
WEIGHT OF ENGINE: Tons ..	18·5	41·9	75·12	79·85	**89**
RATIO: Lbs. Tractive Effort, per ton ..	111·9	230	334	396	**452·8**

Locomotive No. 5955 "Garth Hall"—the first G.W.R. Passenger Engine converted to burn oil fuel

LOOKING AHEAD

ALTHOUGH the earliest G.W.R. locomotives raised steam by the combustion of coke, for more than a century practically all engines have been designed to burn coal, which has been largely drawn from the South Wales coal-fields where the finest steam coal has been available in ample quantity. This fact has considerably influenced the design of Great Western engines, particularly so far as the boiler unit is concerned. Exceptions to the use of coal for propulsion are the diesel-powered rail cars and a diesel-electric shunting locomotive. Incidentally, a number of the latter are now in course of construction.

It will be common knowledge that of late the coal position in this country has caused some anxiety, and the Great Western Railway has turned its attention to the use of fuel oil for firing some of its engines. Up to the time of going to press, ten large freight engines of the 2–8–0 type, and one " Hall " class mixed traffic (4–6–0) engine, have been converted to burn oil fuel. This experiment is to be extended to some engines in the 2–8–2 tank group, and a number of the " Castle " class (4–6–0) are being similarly converted. As a result of this changeover to oil fuel it is expected that a substantial tonnage of coal will be saved annually.

Conversion from coal to oil burning consists in equipping a standard locomotive tender with a tank to carry the oil, and replacing the firegrate and ashpan of the boiler by a special firepan lined with fire brick, the lining being carried over the lower part of the firebox walls. With the exception

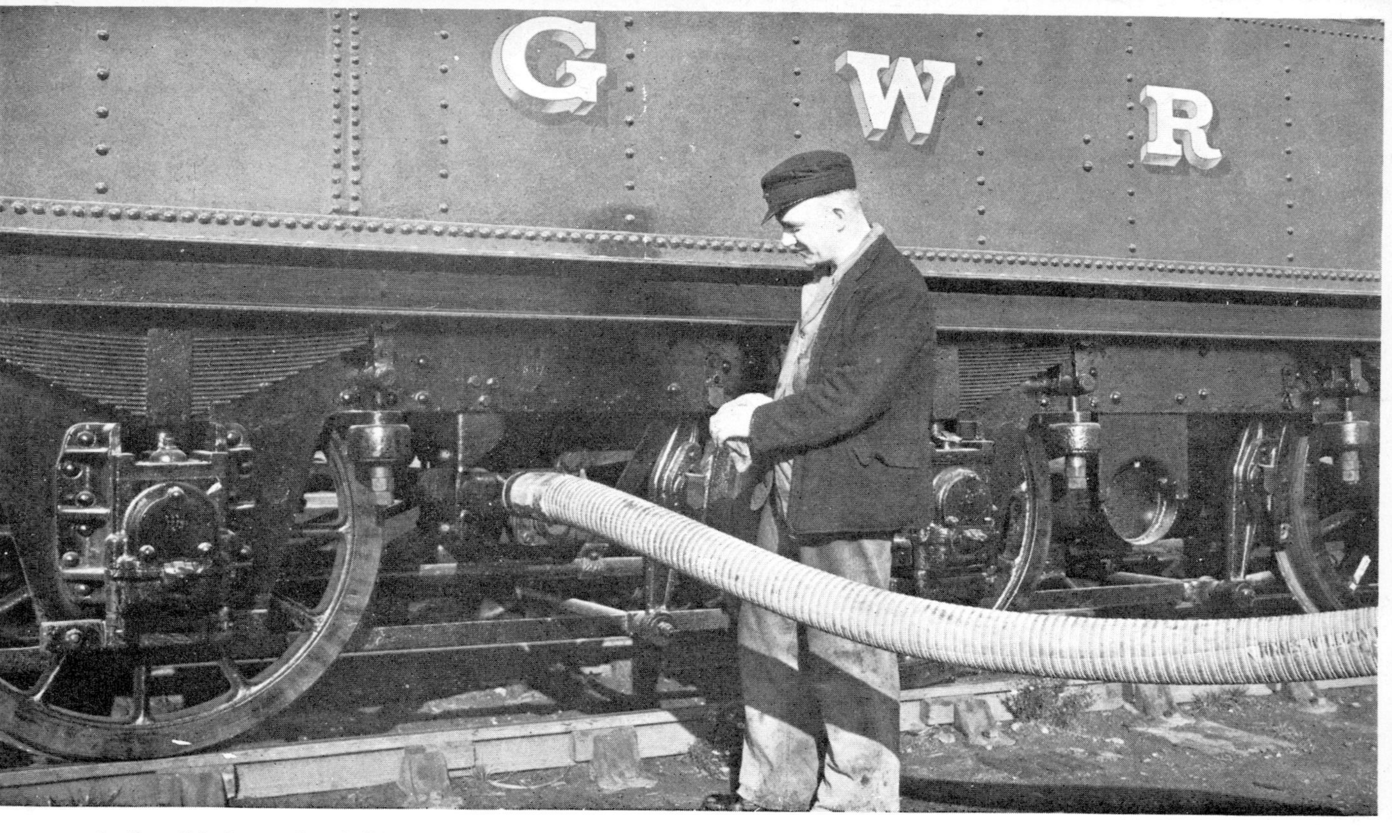

Feeding oil fuel to tender of oil burning locomotive from storage tank

of the oil tank in place of the usual quantity of coal in the tender, little difference is noticeable in the outside appearance of the engine and tender after conversion.

In consistently pursuing its progressive policy in the matter of locomotive practice the Company is now actively engaged in exploring the use of the gas turbine as a power unit for railway locomotives. During the war considerable advances were made in the use of fuel oil for internal combustion engines, and perhaps the apex was achieved in the development of the jet engine for aeroplanes. The principles involved in the design and operation of jet engines are now being applied to the design of a locomotive for main line purposes. This does not mean that trains will be "jet-propelled," but that in the case of a locomotive a gas turbine will drive an air compressor which will feed air into a combustion chamber where the fuel oil is burned under pressure, the heated gases, after passing the turbine, being ejected through louvres in the roof. The gas turbine drives an electric generator which in turn supplies current to traction motors geared to the wheels of the locomotive.

With the coming of the gas turbine locomotive, the familiar shape will disappear. While G.W.R. locomotive enthusiasts may look in vain for pistons, crossheads and connecting rods moving under the engine, and for the familiar white plume at the chimney (if any), they will realise that this is an initial experiment designed to ascertain the relative merits of a new type of power unit compared with the long established use of steam. They will further realise that in this experiment the Great Western Railway is maintaining its reputation for keeping closely in step with modern progress in this scientific age.

G. W. R.

Year-end Locomotive Stock Totals 1938—1945

Type	1938	1939	1940	1941	1942	1943	1944	1945
4–6–0 ..	511	554	569	594	609	621	632	638
4–4–0 ..	105	93	95	95	95	95	94	90
2–8–0 ..	161	173	190	193	226	226	226	226
2–6–0 ..	290	271	274	274	274	274	273	268
2–4–0 ..	3	3	3	3	3	3	3	3
0–6–0 ..	250	253	151	171	171	170	177	175
Total Tender Locos ..	1320	1347	1282	1330	1378	1389	1405	1400
2–8–2T ..	50	54	54	54	54	54	54	54
2–8–0T ..	143	141	151	151	151	151	151	151
2–6–2T ..	428	441	442	442	442	442	442	442
2–4–0T ..	27	27	27	27	26	26	23	16
0–8–2T	1	1	1	1	1	1	1	1
0–6–2T ..	407	405	405	405	405	405	405	402
0–6–0T ..	1123	1153	1195	1214	1243	1271	1281	1269
0–4–2T ..	111	111	111	111	109	109	108	104
0–4–0T ..	20	20	20	20	19	19	19	19
Total Tank Locos. ..	2310	2353	2406	2425	2450	2478	2484	2458
Total Steam Locos. ..	3630	3700	3688	3755	3828	3867	3889	3858
Petrol—Oil ..	1	1	1	1	1	1	1	1
Service Locos. ..	6	4	4	4	4	4	4	4
Rail Motors Petrol—Oil	18	18	32	36	38	38	38	38

G.W.R. LOCOMOTIVE DEPOTS

IT is evident that many G.W.R. locomotive enthusiasts not only take special interest in particular engines, but also in the changes in their respective home depots.

For their information, the *Great Western Railway Magazine* periodically publishes details of changes in the allocation of engines to depots, as well as condemnations and new engines passed into traffic.

The following list sets out the depots to which engines are allocated in each of the Divisions, with the numbers and the code letters stencilled on the engines to indicate their home depots.

List of Locomotive Depots on the Great Western Railway to which Engines are allocated.

	No.	Code letters
London Division		
Didcot	41	DID
Old Oak Common	101	PDN
Oxford	111	OXF
Reading	121	RDG
Southall	131	SHL
Slough	141	SLO
Bristol Division		
Bristol (Bath Road)	22	BRD
Bristol (St. Philip's Marsh) ..	32	SPM
Swindon	132	SDN
Weymouth	162	WEY
Westbury	172	WES
Yeovil	192	YEO
Newton Abbot Division		
Exeter	53	EXE
Laira	83	LA
Newton Abbot	133	NA
Penzance	153	PZ
St. Blazey	173	SBZ
Taunton	203	TN
Truro	213	TR
Wolverhampton Division		
Birkenhead	24	BHD
Banbury	44	BAN
Chester	54	CHR
Croes Newydd	64	CNYD
Crewe	74	CRW
Leamington	94	LMTN
Oxley..	114	OXY
Shrewsbury	134	SALOP
Stourbridge	154	STB

List of Locomotive Depots on the Great Western Railway to which Engines are allocated—*continued.*

Wolverhampton Division (continued)	*No.*	*Code letters*
Tyseley	174	TYS
Wellington	184	WLN
Wolverhampton	194	SRD
Worcester Division		
Cheltenham	45	CHEL
Gloucester	85	GLO
Hereford	95	HFD
Kidderminster	125	KDR
Lydney	175	LYD
Worcester	215	WOS
Newport Division		
Aberbeeg	16	ABEEG
Aberdare	26	ABDR
Cardiff	66	CDF
Ebbw Junction	76	NPT
Llantrisant	86	LTS
Newport Pill	106	PILL
Pontypool Road	126	PPRD
Severn Tunnel Junction	136	STJ
Tondu	146	TDU
Neath Division		
Carmarthen	37	CARM
Danygraig	47	DG
Duffryn Yard (Port Talbot) ..	57	DYD
Goodwick	77	FGD
Landore	87	LDR
Llanelly	107	LLY
Neath	137	NEA
Neyland	167	NEY
Swansea East Dock	197	SED
Whitland	217	WTD
Cardiff Valleys Division		
Abercynon	18	C.V.AYN
Barry..	28	C.V.BRY
Cae Harris	38	CH
Cardiff East Dock	48	CED
Cathays	58	CHYS
Merthyr	68	MTHR
Ferndale	88	C.V.FDL
Radyr	98	RYR
Rhymney	108	RHY
Treherbert	128	THT
Central Wales Division		
Brecon	39	BCN
Machynlleth	119	MCH
Oswestry	129	OSW

ALPHABETICAL INDEX TO NAMED ENGINES

The names of odd engines *not* in standard groups (generally from absorbed railways) are printed in italics.

Engine names marked (*) have been allotted, but the engines (in course of construction) are not in service at the time of publication of this list.

Name.	No.	Page
Abberley Hall ..	4981	43
Abbotsbury Castle ..	4083	55
Abergavenny Castle	5013	55
Aberporth Grange ..	6860	47
Aberystwyth Castle	4084	55
Abney Hall	6900	44
Acton Hall	4982	43
Adderley Hall ..	4901	43
Albert Hall	4983	43
Albrighton Hall ..	4984	43
Aldborough Hall ..	6931	44
Aldenham Hall ..	4902	43
Aldersey Hall ..	6930	44
Alfred Baldwin ..	3363	38
Allesley Hall ..	4985	43
Anthony Manor ..	7801	49
Arbury Hall ..	5986	44
Arley Hall	6901	44
Arlington Court ..	2931	41
Arlington Grange ..	6800	47
Ashburnham ..	2192	—
Ashford Hall ..	5966	44
Ashley Grange ..	6824	47
Ashton Court ..	2932	41
Ashwicke Hall ..	5976	44
Astley Hall	4903	43
Aston Hall	4986	43
Australia	3393	38
Aylburton Grange ..	6801	47
Baglan Hall	4913	43
Bampton Grange ..	6802	47
Barbury Castle ..	5095	55
Barcote Manor ..	7803	49
Barningham Hall ..	6920	44
Barton Hall	4905	43
Bath Abbey ..	5083	55
Baydon Manor ..	7804	49
Beachamwell Hall ..	6934	44
Bearley Grange ..	6831	47
Beaufort	5078	55
Beckford Hall ..	5977	44
Beenham Grange ..	6808	47
Belmont Hall ..	6903	44
Berkeley Castle ..	4085	55
Berrington Hall ..	4912	43
Berry Pomeroy Castle	5012	55
Beverston Castle ..	5068	55
Bibury Court ..	2933	41
Bickmarsh Hall ..	5967	44
Bingley Hall ..	5921	43
Binnegar Hall ..	4904	43
Birchwood Grange ..	6807	47
Birtles Hall	6933	44
Bishop's Castle ..	5064	55
Blackbird	3441	38
Blackwell Grange ..	6806	47
Blaisdon Hall ..	4910	43

Alphabetical Index to Named Engines—*continued.*

Name.	No.	Page	Name.	No.	Page
Blakemere Grange ..	6810	47	Butleigh Court ..	2934	41
Blakesley Hall ..	4909	43	Butlers Hall ..	6902	44
Blasius	3341	38			
Blenheim	5073	55			
Bodicote Grange ..	6870	47			
Bodinnick Hall ..	5978	44			
Bombay	3408	38			
Borwick Hall ..	6921	44			
Bostock Hall ..	5988	44			
Bourton Grange ..	6871	47			
Bradfield Hall ..	4906	43			
Bradley Manor ..	7802	49	Caerphilly Castle ..	4073	55
Breccles Hall ..	6936	44	Calcot Grange ..	6833	47
Brecon Castle ..	5023	55	Calcutta	3406	38
Bride of Lammermoor	2987	41	Caldicot Castle ..	4074	55
Bridgwater Castle	5096	55	Calveley Hall ..	6939	44
British Monarch ..	4021	53	Campion Hall ..	5941	44
Brocket Hall ..	5987	44	Caradoc Grange ..	6873	47
Brockington Grange	6804	47	Cardiff Castle ..	4075	55
Brockley Hall ..	4987	43	Cardigan Castle ..	4087	55
Brockton Grange ..	6832	47	Carew Castle ..	5024	55
Broome Hall ..	4908	43	Carmarthen Castle ..	4076	55
Broome Manor ..	7805	49	Caxton Hall ..	5922	43
Broughton Castle ..	5033	55	Caynham Court ..	2935	41
Broughton Grange ..	6805	47	Cefntilla Court ..	2936	41
Broughton Hall ..	4907	43	Chaffinch	3443	38
Browsholme Hall ..	6935	44	Charfield Hall ..	6904	44
Buckenhill Grange ..	6830	47	Charles J. Hambro ..	2978	41
Bucklebury Grange..	6803	47	Chepstow Castle ..	4077	55
Builth Castle ..	4086	55	Cherwell Hall ..	4989	43
Bullfinch	3442	38	Chesford Grange ..	6812	47
Bulwell Hall.. ..	4988	43	Chicheley Hall ..	6906	44
Burghclere Grange ..	6809	47	Childrey Manor ..	7809	49
Burmington Grange	6829	47	Chirk Castle ..	5025	55
Burry Port	2193	—	Claughton Hall ..	6905	44
Burton Hall ..	6922	44	Cleeve Abbey ..	5091	55
Burwarton Hall ..	6932	44	Cleeve Grange ..	6850	47

Alphabetical Index to Named Engines—*continued.*

Name.	No.	Page	Name.	No.	Page
Clevedon Court ..	2937	41	County of Hereford	1017	51
Clifford Castle* ..	5098	55	County of Leicester	1018	51
Clifton Hall	4990	43	County of Merioneth	1019	51
Clyffe Hall	5951	44	County of Middlesex	1000	51
Cobham Hall ..	4991	43	County of Monmouth*	1020	51
Cockington Manor ..	7806	49	County of Montgomery* ..	1021	51
Cœur de Lion ..	2980	41	County of Northampton* ..	1022	51
Cogan Hall	5952	44	County of Oxford* ..	1023	51
Coity Castle	5035	55	County of Pembroke*	1024	51
Colston Hall ..	5923	43	County of Radnor*	1025	51
Comet	3283	—	County of Salop* ..	1026	51
Compton Castle ..	5099	55	County of Somerset	1004	51
Compton Manor ..	7807	49	County of Stafford*	1027	51
Condover Hall ..	4915	43	County of Warwick*	1028	51
Cookham Manor ..	7808	49	County of Wilts ..	1003	51
Corfe Castle	5034	55	County of Worcester*	1029	51
Conyngham Hall ..	6937	44	Cranbourne Grange..	6811	47
Corndean Hall ..	6938	44	Cranmore Hall ..	4914	43
Cormorant	3444	38	Cransley Hall ..	5989	44
Cornubia†	3254	—	Crawley Grange ..	6872	47
Corsham Court ..	2938	41	Criccieth Castle ..	5026	55
Cory Hall	5968	44	Croome Court ..	2939	41
Countess	823	—	Crosby Hall	4992	43
County of Berks ..	1002	51	Crosswood Hall ..	4917	43
County of Brecknock	1007	51	Croxteth Hall ..	6923	44
County of Bucks ..	1001	51	Cruckton Hall ..	5979	44
County of Cardigan	1008	51	Crumlin Hall ..	4916	43
County of Carmarthen	1009	51	Crynant Grange ..	6861	47
County of Carnarvon	1010	51			
County of Chester ..	1011	51			
County of Cornwall..	1006	51			
County of Denbigh	1012	51			
County of Devon ..	1005	51			
County of Dorset ..	1013	51			
County of Glamorgan	1014	51			
County of Gloucester	1015	51			
County of Hants ..	1016	51			

† "Duke" (3252) class still in service.

Alphabetical Index to Named Engines—*continued.*

Name.	No.	Page	Name.	No.	Page
Dalton Hall	4993	43	Earl Baldwin ..	5063	55
Dartington Hall ..	4918	43	Earl Bathurst ..	5051	55
Dartmouth Castle ..	4088	55	Earl Cairns	5053	55
Davenham Hall ..	6907	44	Earl Cawdor.. ..	5046	55
Defiant	5080	55	Earl of Berkeley ..	5060	55
Denbigh Castle ..	7001	55	Earl of Birkenhead	5061	55
Derwent Grange ..	6862	47	Earl of Clancarty ..	5058	55
Devizes Castle ..	7002	55	Earl of Dartmouth ..	5047	55
Didlington Hall ..	6940	44	Earl of Devon ..	5048	55
Dingley Hall ..	5980	44	Earl of Ducie ..	5054	55
Dinton Hall	5924	43	Earl of Dunraven ..	5044	55
Doldowlod Hall ..	5942	44	Earl of Dudley ..	5045	55
Dolhywel Grange ..	6863	47	Earl of Eldon ..	5055	55
Dorford Hall ..	5990	44	Earl of Mount Edgcumbe	5043	55
Dominion of Canada	3391	38	Earl of Plymouth ..	5049	55
Donnington Castle ..	4089	55	Earl of Powis ..	5056	55
Donnington Hall ..	4919	43	Earl of Radnor ..	5052	55
Dorchester Castle ..	4090	55	Earl of St. Germans	5050	55
Dorney Court ..	2940	41	Earl of Shaftesbury	5062	55
Downham Hall ..	6908	44	Earl St. Aldwyn ..	5059	55
Downton Hall ..	4994	43	Earl Waldegrave ..	5057	55
Draycott Manor ..	7810	49	Earlestoke Manor ..	7812	49
Dudley Castle ..	4091	55	Eastbury Grange ..	6813	47
Dumbleton Hall ..	4920	43	Eastcote Hall ..	5925	43
Dummer Grange ..	6834	47	Eastham Grange ..	6835	47
Dunley Hall ..	5953	44	Eastnor Castle ..	7004	55
Dunley Manor ..	7811	49	Easton Court ..	2941	41
Dunraven Castle ..	4092	55	Easton Hall	4995	43
Dunster Castle ..	4093	55	Eaton Hall	4921	43
Dymock Grange ..	6864	47	Eden Hall	4996	43
Dynevor Castle ..	4094	55	Elmdon Hall ..	5943	44
			Elmley Castle ..	7003	55
			Elton Hall	4997	43
			Enborne Grange ..	6814	47
			Enville Hall ..	4922	43
			Eshton Hall ..	6942	44

Alphabetical Index to Named Engines—*continued.*

Name.	No.	Page	Name.	No.	Page
Estevarney Grange	6836	47	Gatacre Hall ..	4928	43
Evenley Hall ..	4923	43	Gladiator	5076	55
Evesham Abbey ..	5085	55	Glasfryn Hall ..	6945	44
Eydon Hall	4924	43	Glastonbury Abbey	4061	53
Eynsham Hall ..	4925	43	Goldfinch	3446	38
Eyton Hall	4998	43	Goodmoor Grange ..	6838	47
			Goodrich Castle ..	5014	55
			Gopsal Hall	4999	43
			Gossington Hall ..	6910	44
			Goytrey Hall ..	4929	43
			Grantley Hall ..	6924	44
			Granville Manor ..	7818	49
			Gresham Hall ..	5991	44
			Grotrian Hall ..	5926	43
			Guild Hall	5927	43
Faendre Hall ..	5954	44	Guinevere†	3256	—
Fairey Battle ..	5077	55	Gwenddwr Grange ..	6817	47
Fairleigh Hall ..	4926	43	*Gwendraeth*	2196	—
Farleigh Castle ..	5027	55			
Farnborough Hall ..	4927	43			
Farnley Hall ..	6943	44			
Fawley Court ..	2942	41			
Fillongley Hall ..	6941	44			
Flamingo	3445	38			
Fledborough Hall ..	6944	44			
Forthampton Grange	6837	47	Haberfield Hall ..	6949	44
Frank Bibby ..	3364	38	Hackness Hall ..	6925	44
Frankton Grange ..	6816	47	Haddon Hall ..	5928	43
Frensham Hall ..	5981	44	Hagley Hall ..	4930	43
Freshford Manor ..	7813	49	Hampton	5074	55
Frewin Hall	6909	44	Hampden Court ..	2943	41
Frilford Grange ..	6815	47	Hanbury Hall ..	4931	43
Frilsham Manor ..	7816	49	Hanham Hall ..	5929	43
Fringford Manor ..	7814	49	Hannington Hall ..	5930	43
Fritwell Manor ..	7815	49	Hardwick Grange ..	6818	47
Garsington Manor ..	7817	49	Harlech Castle ..	4095	55
Garth Hall	5955	44	Harrington Hall ..	5982	44

† "Duke" (3252) class still in service

Alphabetical Index to Named Engines—*continued.*

Name.	No.	Page
Hatherley Hall	5931	43
Hatherton Hall	4932	43
Haughton Grange	6874	47
Haydon Hall	5932	43
Hazel Hall	5901	43
Hazeley Grange	6840	47
Headbourne Grange	6852	47
Heatherden Hall	6946	44
Helmingham Hall	6947	44
Helmster Hall	6912	44
Hengrave Hall	5970	44
Henley Hall	5983	44
Hewell Grange	6839	47
Highclere Castle	4096	55
Highnam Court	2944	41
Highnam Grange	6819	47
Hilda	359	—
Hillingdon Court	2945	41
Himley Hall	4933	43
Hinderton Hall	5900	43
Hindford Grange	6875	47
Hindlip Hall	4934	43
Hinton Manor	7819	49
Holbrooke Hall	6948	44
Holker Hall	6911	44
Holkham Hall	6926	44
Honington Hall	5969	44
Hopton Grange	6865	47
Horsley Hall	5956	44
Horton Hall	5992	44
Howick Hall	5902	43
Hurricane	5072	55
Hurst Grange	6851	47
Hutton Hall	5957	44
Ickenham Hall	5944	44
Impney Hall	6951	44
Inchcape	3430	38
Isambard Kingdom Brunel	5069	55
Isle of Jersey†	3284	—
Ivanhoe	2981	41
Jackdaw	3447	38
Keele Hall	5903	43
Kelham Hall	5904	43
Kenilworth Castle	4097	55
Ketley Hall	4935	43
Kidwelly	2194	—
Kidwelly Castle	4098	55
Kilgerran Castle	4099	55
Kimberley Hall	6952	44
King Charles I	6010	58
King Charles II	6009	58
King Edward I	6024	58
King Edward II	6023	58
King Edward III	6022	58
King Edward IV	6017	58
King Edward V	6016	58

† "Duke" (3252) class still in service

Alphabetical Index to Named Engines—*continued.*

Name.	No.	Page	Name.	No.	Page
King Edward VI ..	6012	58	Knight of the Grand Cross	4018	53
King Edward VII ..	6001	58	Knight of the Thistle	4012	53
King Edward VIII	6029	58	Knight Templar ..	4019	53
King George I ..	6006	58	Knolton Hall ..	5958	44
King George II ..	6005	58	Knowsley Hall ..	5905	43
King George III ..	6004	58			
King George IV ..	6003	58			
King George V ..	6000	58			
King George VI ..	6028	58			
King Henry III ..	6025	58			
King Henry IV .	6020	58			
King Henry V ..	6019	58	Lady Macbeth ..	2905	41
King Henry VI ..	6018	58	*Lady Margaret* ..	1308	—
King Henry VII ..	6014	58	Lady of Lynn ..	2906	41
King Henry VIII ..	6013	58	Lady of Lyons ..	2903	41
King James I ..	6011	58	Lady of Quality ..	2908	41
King James II ..	6008	58	Lady of the Lake ..	2902	41
King John	6026	58	Lamphey Castle ..	7005	55
King Richard I ..	6027	58	Lanelay Hall ..	4937	43
King Richard II ..	6021	58	Langford Court ..	2946	41
King Richard III ..	6015	58	Langton Hall ..	6914	44
King William III ..	6007	58	Launceston Castle ..	5000	55
King William IV ..	6002	58	Lawton Hall ..	5906	43
Kingfisher	3448	38	Leaton Grange ..	6821	47
Kingsland Grange ..	6876	47	Leckhampton Hall ..	5945	44
Kingsthorpe Hall ..	6950	44	Leighton Hall ..	6953	44
Kingstone Grange ..	6820	47	Levens Hall ..	6913	44
Kinsgway Hall ..	5933	43	Liddington Hall ..	4938	43
Kingswear Castle ..	5015	55	Lilford Hall	6927	44
Kinlet Hall	4936	43	Linden Hall	5984	44
Kirby Hall	6993	44	Littleton Hall ..	4939	43
Kneller Hall ..	5934	43	Llandovery Castle ..	5001	55
Knight Commander	4020	53	Llanfrechfa Grange	6827	47
Knight of Liège ..	4017	53	Llangedwyn Hall ..	4941	43
Knight of St. John	4015	53	Llanstephan Castle	5004	55
Knight of St. Patrick	4013	53	Llanthony Abbey ..	5088	55

Alphabetical Index to Named Engines—*continued.*

Name.	No.	Page	Name.	No.	Page
Llantilio Castle ..	5028	55	Misterton Hall ..	6916	44
Llanvair Grange ..	6825	47	Monmouth Castle ..	5037	55
Lloyd's	100.A1	55	Montgomery Castle	5016	55
Lockheed Hudson ..	5081	55	Morehampton Grange	6853	47
Lode Star	4003	53	Moreton Hall ..	5908	43
Longford Grange ..	6878	47	Morfa Grange ..	6866	47
Lord Mildmay of Flete	3417	38	Morlais Castle ..	5038	55
Lotherton Hall ..	6954	44	Morning Star ..	4004	53
Ludford Hall ..	4940	43	Moseley Hall ..	4946	43
Ludlow Castle ..	5002	55	Mostyn Hall ..	5985	44
Lulworth Castle ..	5003	55	Mottram Hall ..	6956	44
Lydcott Hall ..	6955	44	Mounts Bay† ..	3273	—
Lydford Castle ..	7006	55	Mursley Hall ..	6915	44
Lyonshall Castle ..	5036	55	Mytton Hall ..	5996	44
Lysander	5079	55			
Madras	3407	38	Nanhoran Hall ..	4947	43
Madresfield Court ..	2947	41	Nannerth Grange ..	6826	47
Maindy Hall.. ..	4942	43	Natal Colony ..	3396	38
Malmesbury Abbey..	4062	53	Neath Abbey ..	5090	55
Manorbier Castle ..	5005	55	Newport Castle ..	5065	55
Manton Grange ..	6822	47	Newton Hall ..	5909	43
Marble Hall	5907	43	Nightingale ..	3449	38
Marlas Grange ..	6841	47	Norcliffe Hall ..	6957	44
Marrington Hall ..	4943	43	North Star	4000	55
Marwell Hall ..	5946	44	Northwick Hall ..	4948	43
Mawley Hall ..	5959	44	Norton Hall ..	5935	43
Mercury†	3287	—	Nunhold Grange ..	6842	47
Merevale Hall ..	5971	44	Nunney Castle ..	5029	55
Merlin†	3259	—			
Middleton Hall ..	4944	43			
Milligan Hall ..	4945	43			

† "Duke" (3252) class still in service,

Alphabetical Index to Named Engines—*continued.*

Name.	No.	Page	Name.	No.	Page
Oakley Grange ..	6823	47	Prince Albert ..	4042	53
Oakley Hall	5936	43	Prince George ..	4044	53
Ogmore Castle ..	7007	55	Prince Henry ..	4043	53
Olton Hall	5972	44	Prince John ..	4045	53
Ottawa	3399	38	Prince of Wales ..	4041	53
Overton Grange ..	6879	47	Princess Alexandra	4053	53
Oxburgh Hall ..	6958	44	Princess Alice ..	4050	53
			Princess Augusta ..	4058	53
			Princess Beatrice ..	4052	53
			Princess Charlotte ..	4054	53
			Princess Elizabeth ..	4057	53
			Princess Eugenie ..	4060	53
Packwood Hall ..	4949	43	Princess Helena ..	4051	53
Park Hall	5910	43	Princess Louise ..	4047	53
Patshull Hall ..	4950	43	Princess Margaret ..	4056	53
Paviland Grange ..	6845	47	Princess Mary ..	4046	53
Peacock	3450	38	Princess Maud ..	4049	53
Peatling Hall ..	6959	44	Princess Patricia ..	4059	53
Pelican	3451	38	Princess Sophia ..	4055	53
Pembroke Castle ..	4078	55	Princess Victoria ..	4048	53
Pendeford Hall ..	4951	43	Priory Hall	4958	43
Pendennis Castle ..	4079	55	Purley Hall	4959	43
Penguin	3452	38	Pyle Hall	4960	43
Penhydd Grange ..	6844	47	Pyrland Hall ..	4961	43
Penrhos Grange ..	6868	47			
Peplow Hall ..	4952	43			
Pershore Plum ..	3353	38			
Peterston Grange ..	6867	47			
Pioneer	2197	—	Queen Adelaide ..	4034	53
Pitchford Hall ..	4953	43	Queen Alexandra ..	4032	55
Plaish Hall	4954	43	Queen Berengaria ..	4038	53
Plaspower Hall ..	4955	43	Queen Boadicea ..	4040	53
Plowden Hall ..	4956	43	Queen Charlotte ..	4035	53
Postlip Hall	4957	43	Queen Elizabeth ..	4036	53
Poulton Grange ..	6843	47	Queen Mary ..	4031	53
Powderham Castle ..	4080	55	Queen Matilda ..	4039	53
Preston Hall ..	5911	43	Queen Victoria ..	4033	53

Alphabetical Index to Named Engines—*continued.*

Name.	No.	Page	Name.	No.	Page
Queen's Hall ..	5912	43	Saint Andrew ..	2913	41
Quentin Durward ..	2979	41	Saint Bartholomew	2915	41
			Saint Benedict ..	2916	41
			Saint Benet's Hall ..	5947	44
			Saint Bride's Hall ..	4972	43
			Saint David ..	2920	41
			St. Donats Castle ..	5017	55
			Saint Edmund Hall	5960	44
Raglan Castle ..	5008	55	St. Fagans Castle ..	5067	55
Ragley Hall ..	4962	43	Saint Helena ..	2924	41
Raveningham Hall ..	6960	44	Saint Martin ..	4900	43
Reading Abbey ..	5084	55	St. Mawes Castle ..	5018	55
Resolven Grange ..	6869	47	Saint Nicholas ..	2926	41
Restormel Castle ..	5010	55	Saint Patrick ..	2927	41
Rhuddlan Castle ..	5039	55	Saint Sebastian ..	2928	41
Rignall Hall ..	4963	43	Saint Stephen ..	2929	41
Ripon Hall	5914	43	Saint Vincent ..	2930	41
River Fal	3379	38	Sarum Castle ..	5097	55
River Plym	3376	38	Seagull	3453	38
Rob Roy	2988	41	Shakenhurst Hall ..	4966	43
Rodwell Hall ..	4964	43	Shirburn Castle ..	5030	55
Rolleston Hall ..	5973	44	Shirenewton Hall ..	4967	43
Rood Ashton Hall ..	4965	43	Shotton Hall ..	4968	43
Rougemont Castle ..	5007	55	Shrewsbury Castle ..	5009	55
Roundhill Grange ..	6854	47	Shrugborough Hall	4969	43
Roydon Hall ..	5994	44	Siddington Hall ..	5948	44
Ruckley Grange ..	6846	47	Sir Arthur Yorke ..	3418	38
Rushton Hall ..	5913	43	Sir Daniel Gooch ..	5070	55
			Sir Watkin Wynn ..	3375	38
			Sketty Hall	4970	43
			Skylark	3454	38
			Soughton Hall ..	6962	44
			Sparkford Hall ..	5997	44
			Spitfire	5071	55
			Stackpole Court ..	2948	41
Saighton Grange ..	6855	47	Stanford Court ..	2949	41
Saint Ambrose ..	2912	41	Stanford Hall ..	5937	43

Alphabetical Index to Named Engines—*continued.*

Name.	No.	Page
Stanley Hall ..	5938	43
Stanway Hall ..	4971	43
Starling	3455	38
Stedham Hall ..	6961	44
Stokesay Castle ..	5040	55
Stowe Grange ..	6856	47
Swallowfield Park ..	4007	53
Sweeney Hall ..	4973	43
Swordfish	5082	55
Talgarth Hall ..	4974	43
Talisman	2989	41
Tangley Hall ..	5939	44
Taplow Court ..	2950	41
Tasmania	3395	38
Tawstock Court ..	2951	41
Thames†	3291	—
The Earl	822	—
The Somerset Light Infantry (Prince Albert's)	4016	55
The South Wales Borderers	4037	55
Thirlstaine Hall ..	6965	44
Thornbridge Hall ..	6964	44
Throwley Hall ..	6963	44
Tidmarsh Grange ..	6847	47
Tintagel Castle ..	5011	55
Tintern Abbey ..	5087	55
Titley Court ..	2953	41
Tiverton Castle ..	5041	55
Tockenham Court ..	2954	41
Toddington Grange	6848	47
Torquay Manor ..	7800	49
Tortworth Court ..	2955	41
Totnes Castle ..	5031	55
Toynbee Hall ..	5961	44
Treago Castle ..	5019	55
Tregenna Castle ..	5006	55
Trellech Grange ..	6828	47
Trematon Castle ..	5020	55
Trematon Hall ..	5949	44
Trentham Hall ..	5915	43
Tre Pol and Pen† ..	3265	—
Tresco Abbey ..	5092	55
Tretower Castle ..	5094	55
Trevithick†	3264	—
Trevor Hall	5998	44
Trinity Hall ..	5916	43
Tudor Grange ..	6857	47
Twineham Court ..	2952	41
Tylney Hall ..	6919	44
Umberslade Hall ..	4975	43
Underley Hall ..	6928	44
Upton Castle ..	5093	55
Usk Castle	5032	55

† "Duke" (3252) class still in service.

Alphabetical Index to Named Engines—*continued.*